AF328574

ADVANCES IN
BIOSENSORS

Volume 3 • 1995

BIOSENSORS: A
RUSSIAN PERSPECTIVE

ADVANCES IN BIOSENSORS

BIOSENSORS: A RUSSIAN PERSPECTIVE

Editors: ANTHONY P. F. TURNER
Cranfield Biotechnology Centre
Cranfield University
Bedfordshire, England

YU. M. YEVDOKIMOV
W.A. Engelhardt Institute of Molecular Biology
The Russian Academy of Sciences
Moscow

VOLUME 3 • 1995

Greenwich, Connecticut   *London, England*

CONTENTS

LIST OF CONTRIBUTORS

A. A. Bayev

W. A. Engelhardt Institute of Molecular
Biology
The Russian Academy of Sciences
Moscow, Russia

V. A. Bogdanovskaya

A. N. Frumkin Institute of
Electrochemistry
The Russian Academy of Sciences
Moscow, Russia

B. A. Chernuha

W. A. Engelhardt Institute of Molecular
Biology
The Russian Academy of Sciences
Moscow, Russia

A. M. Egorov

Division for Chemical Enzymology
Department of Chemistry
M. V. Lomonosov State University
Moscow, Russia

S. M. Khomutov

Biosensors Group
Moscow Institute of Radioengineering
Moscow, Russia

B. B. Kim

Division for Chemical Enzymology
Department of Chemistry
M. V. Lomonosov State University
Moscow, Russia

M. P. Kirpichnikov

W. A. Engelhardt Institute of Molecular
Biology
The Russian Academy of Sciences
Moscow, Russia

A. A. Kononenko

Faculty of Biology
Moscow State University
Moscow, Russia

I. N. Kurochkin

Department of Biosensors
Research Center of Molecular Diagnostics
and Therapy
Moscow, Russia

B. A. Kuznetsov

A. N. Bach Institute of Biochemistry
The Russian Academy of Sciences
Moscow, Russia

E. P. Lukashev

Faculty of Biology
Moscow State University
Moscow, Russia

A. P. Osipov

Division for Chemical Enzymology
Department of Chemistry
M. V. Lomonosov State University
Moscow, Russia

D. B. Papkovsky

Department of Biosensors
Research Center of Molecular Diagnostics
and Therapy
Moscow, Russia

A. N. Reshetilov

Biosensors Group
Moscow Institute of Radioengineering
Moscow, Russia

A. P. Savitsky

A. N. Bach Institute of Biochemistry
The Russian Academy of Sciences
Moscow, Russia

V. V. Savransky

A. N. Bach Institute of Biochemistry
The Russian Academy of Sciences
Moscow, Russia

S. G. Skuridin

W. A. Engelhardt Institute of
Molecular Biology
The Russian Academy of Sciences
Moscow, Russia

M. R. Tarasevich

A. N. Frumkin Institute of Electrochemistry
The Russian Academy of Sciences
Moscow, Russia

A. I. Yaropolov

A. N. Bach Institute of Biochemistry
The Russian Academy of Sciences
Moscow, Russia

Y. M. Yevdokimov

W. A. Engelhardt Institute of Molecular
Biology
The Russian Academy of Sciences
Moscow, Russia

PREFACE

BIOSENSORS: A RUSSIAN PERSPECTIVE

Biosensors are products of a newly emerging area of multidisciplinary technologies, which spans biotechnology, molecular biology, material science and microelectronics.

What is a biosensor? Using various definitions of biosensors we can compile one, which emphasises the idea of the contributions from very different branches of science. "A biosensor is an analytical tool or system consisting of an immobilized biological material which converts information about the properties of the analyte(s) into a quantifiable signal via a suitable transducer." This information, in a trivial sense, can be given in the form of different factors, such as, for example, concentrations of biologically active or toxic compounds. Specific data may also be derived, however, such as the concentration of a particular ion or chemical in a medium. The signal generated in a system can involve a variety of physical parameters, but is normally converted to a digital electronic form.

The figure concluding this preface shows a generalized scheme for all types of biosensors. It illustrates the main concept of the biosensor: "Biological molecules, coupled to a transducer, translate biological 'activity' directly into measured electric current." The scheme illustrates that only one component (dark triangles) of the mixture interacts specifically with the biomolecules. The generated signal is transformed by means of a transducer, enhanced and processed using a microprocessor to produce data on the properties of the mixture under study.

Enzymes, nucleic acids, antibodies, receptors, membranes, intact cells, and tissue pieces have been used for fabrication of biosensors. The most important part of any biosensor is this biological sensing element (also called a biospecific surface or biosensing unit). The term "biosensing" denotes acquisition of information by biomolecular system. The biosensing unit may be considered as an ensemble of biological molecules that utilize events at molecular level to detect chemical and biochemical substances in a form of a characteristic signal. Physico-chemical events that result from a reaction between the molecules forming the biosensing unit and molecules in the analyzed medium determine the specificity and the efficiency of a biosensor.

Theoretically, the sensitivity of a biosensor is related to a single biopolymeric molecule. In practice, however, it depends on the sensitivity of the detection system, which in turn is connected with the density of location of biopolymeric molecules on a biosensing surface. To reach a high extent of ordering of biopolymeric molecules different methods of immobilization are used. Immobilization facilitates the formation of a layer of biopolymeric molecules or film containing these molecules. Immobilization of biomolecules allows the required density to be achieved while retaining as much of the native properties of the biopolymeric molecules as possible. Optimization of immobilization is a key issue and sometimes addition of agents such as antioxidants, cofactors and detergents is also necessary.

Immobilization of biopolymeric molecules opens up the possibility of creating biosensors which differ in "operating principle:" (1) biosensors of classical type, where immobilized biopolymeric molecules contact directly with a transducer, or (2) biosensors of a disposable type when a film containing immobilized biopolymeric molecules is removed from the transducer.

Since "biosensing" is a process based on the principle of "molecular recognition" combined with physicochemical events that accompany "recognition," it follows that there are basically two types of biosensors.

The first type are "bioaffinity" biosensors. Here immobilized biopolymeric molecules "recognize" the compounds that are presented in the solution, discriminate between different compounds, and detect only one. The formation of a complex with immobilized biomolecules is the basic mechanism which occurs in this type of biosensor. The properties of the complex formed (shape, color, etc.) are different from those of the initial molecules. Such differences constitute a "signal" for the system. The signal is amplified electronically by an appropriate transducer, and is proportional to the concentration of the compound analyzed.

The second basic type is the "enzymatic/metabolic" biosensor. Here, the molecules of an immobilized enzyme "recognize" substrate molecules contained in a solution. The product (e.g. protons, ions) released as a result of the biologically catalyzed reaction between the incoming agent (substrate) and the immobilized enzyme is the "signal" for the system, and its magnitude is proportional to the concentration of the analyte. This type of biosensor is already available commercially in the form of "enzyme electrodes."

Receptor–ligand-or enzyme–substrate-based biosensors, which "recognize" the presence of ligands with similar properties, are suitable for detection of groups of analytes; this is so-called "group or class determination." "Group determination" allows rapid detection of the compounds of interest without identifying a specific agent.

The immediate advantages offered by biosensors include:

- specificity of analysis even in complex mixtures, negating the necessity for pretreatment of specimens being studied;
- analysis of small volumes of specimens in a short period within a real-time scale;
- potential for provision of on-line analysis and therefore feedback control, made possible by the compatibility of biosensors with microprocessors;
- unskilled personnel can perform the analysis after a minimum of training; and
- low cost of biosensor production and instrumentation.

Biosensors are small, simple to use, portable, and produce data that can be channeled directly into computer analysis. There are examples of biosensors which can detect particular classes of substances in seconds, detect as little as 10^{-21} moles, are stable in use for 3 months or more, and stable on storage for at least 18 months.

Creating biosensors, in essence, involves the resolution of two problems which are related to different fields of science. First, a specific biosensing unit must be fabricated which "recognizes" biologically active compounds with the highest efficiency. This can be solved within the biological sciences. Second, an adequate recording scheme for detection of the signal which appears in the system must be created. This should be solved within the technical sciences.

Since the first commercially successful biosensors appeared in 1975, interest in the area has grown steadily. Approximately 1000 publications on biosensors appeared last year and the number is steadily growing.

Numerous meetings and workshops on biosensors have been held and biosensors are now applied in various fields of science and industry. The world market for biosensors is currently approximately $240 million and is expected to reach $750 million by the year 2000. It is reported that research on biosensors is being conducted by over 50 companies in Japan alone. Major U.S. companies with advanced research programs include Abbott Laboratories, Du Pont, Corning, Honeywell, and Miles Laboratories. New biosensors are being developed in the U.S.A. by specialized bioengineering companies such as Molecular Devices, Yellow Spring Instrument Company, i-STAT, and Medisense. Internationally, over 500 companies and research groups are thought to be active in the area.

This volume provides a unique perspective of biosensor activity by focusing on the various types of biosensor research and the practical application of biosensors achieved by research groups exclusively from Russian scientific organizations.

The appearance of such a geographically specialized volume perhaps needs some further explanation. One evening in August 1992, a group of scientists from various countries was invited by the organizers of FEBS (Federation of European Biochemical Societies) to test "Guinness" in a long favorite inner city Dublin restaurant, the "Old Dublin." The Guinness was excellent —talks about life, politics and science were stimulated by the friendly atmosphere of this meeting. For a short while the conversation focused on the problems of "new Russia." Discussions demonstrated that receiving and exchanging scientific information from a new Russia is still a problem, especially in the case of quickly developing branches of science. However, some Russian papers slowly trickling through the system clearly contain important scientific approaches. Because all the members of the group around the table had a commitment to biosensors, a hazy, but interesting idea was hatched: to publish a volume of *Advances in Biosensors* with contributors exclusively from Russia! The appearance of this volume is a "reply" to an international challenge. It shows that scientists from the U.K. and Russia can easily realize their aspiration to cooperate.

The selection of the contributions was performed in conjunction with a scientific council for the area of "Biosensors" within the framework of the Russian State Program, "Newest Methods of Bioengineering." We, the editors, are indebted to our colleagues from Dublin City University, Professor M. Smith and Professor R. O'Kennedy, for organizing the meeting between "biosensorists" from different countries and for their hospitality, which has catalyzed our collaboration. We also express our gratitude to all the Russian contributors who supported the idea to present new results, that, in the main, are unknown to a wide audience. We hope that this volume will stimulate international contacts as well as scientific competition in the field of biosensors.

Anthony P. F. Turner
Yu. M. Yevdokimov
Editors

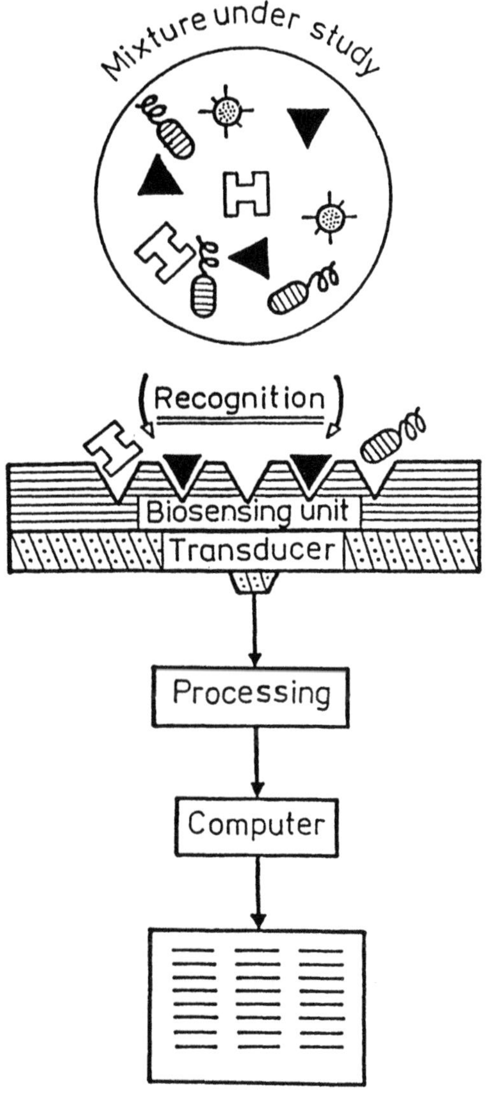

Numerical information
about properties of studed mixture

A General Scheme for Biosensor Devices

BIOSENSER RESEARCH IN RUSSIA

A. A. Bayev and M. P. Kirpichnikov

I. INTRODUCTION

The discoveries in the basic life sciences during the past 20 years are of great importance as components of progress in research and engineering, and they have demonstrated clearly their tremendous significance to mankind. Together with giant advances in understanding the vital function mechanism and principles of the living organism and its interrelationship with the environment, a "new biotechnology" has arisen based on achievements of physicochemical biology, genetics, and also genetic and cell engineering. Under increasing anthropogenic loads on one hand, and an increasing level of living standards on the other, providing the full value of nutrition and health for humans, solving ecological

Advances in Biosensors
Volume 3, pages 1-4.
Copyright © 1995 by JAI Press Inc.

ISBN:1-55938-535-9

problems, and creating new industrial processes are impossible without new developments in biotechnology.

The comprehension of these facts has resulted in a change of priorities in the stated scientific-industrial policy of developed countries all over the world in favor of life sciences and new biotechnology. One example of modern biotechnology is the creation of different biologically important macromolecules or their complexes for use as biosensors in various fields of industry as well as in environmental control. The advantages of biosensors are their high specificity and sensitivity, the high rate and simplicity of analysis, the possibility of coupling with computers, and the low cost per analysis.

Biosensors in the form of so-called "enzyme electrodes" have been commercially available since 1975. During the coming years commercial realization will be concerned mainly with biosensors for analysis of low molecular mass compounds such as glucose, lactate, alcohol, cholesterol, urea, uric acid, and others. According to our information biosensors are used in Western countries in medicine, environmental control the food industry and control of industrial processes.

Despite a high level of activity in the fields of biotechnology, physical chemistry of biological macromolecules and molecular biology in the former USSR, experimentation on biosensors was rather incomplete. Investigations on biosensors were carried out mainly at the Institutes of the USSR Academy of Sciences and the Ministry of Medical Industry of the USSR. But many of these studies, with the exception of a few notable items, did not become public knowledge. Among the early works, the studies of Prof. Yu. Kulis (Lithuania) on the creation of a glucose biosensor should be mentioned.

2.   COORDINATED RESEARCH

In 1987 at the seminar organized in Pushchino-on-Oka by the State Committee for Science and Technology (GKNT) and the Academy of Sciences of the USSR, Prof. A.A. Bayev brought up the question of coordination of the Soviet researchers dealing with biosensor studies. About 50 researchers from the scientific centers of Russia, Ukraine, Belorussia, Armenia, and Lithuania, took part in the seminar. It was considered necessary to formulate a joint scientific program for investigations in field of biosensors.

However, this decision was only realized in 1990. The program on biosensors was developed as an independent part of the branch "Engineering Enzymology" in the State Research Program, "Newest Methods of Bioengineering."

This State Program brought together the efforts of such scientific centers as the Institute of Biochemistry of the Lithuanian Academy of Sciences, Research enterprise "Fermentas" (Vilnius), M.V. Lomonosov State University, N.A. Bach Institute of Biochemistry of the USSR Academy of Sciences, W.A. Engelhardt

Institute of Molecular Biology of the USSR Academy of Sciences, the Institute of Electrochemistry of the USSR Academy of Sciences, the All-Union Research Centre of Molecular Diagnostics and Treatment (Ministry of Health of the USSR), the Institute of General Physics of the USSR Academy of Sciences, the Institute of Analytical Chemistry and Geochemistry of the USSR Academy of Sciences, the All-Union Research Institute of Biological Equipment Building of the Concern "Biopreparat" (all in Moscow), the Physical Institute (Yerevan, Armenia), Kiev State University, the Institute of Molecular Biology and Genetics of the Ukrainian Academy of Sciences, Research Enterprise "Microprocessor" (Ministry of Electronic Industry of the USSR) (all in Kiev), the Experimental Construction Bureau for Fine Biological Machine Building (Kirishi) and some others. The new program also involved industries such as the Production Association "Sigma" (Vilnius), the joint venture "BioChimMac" (Moscow) and small ventures such as "Immunotech" and "Bioreactor" (Moscow).

The disintegration of the USSR resulted in the breaking of the scientific contacts between the research groups from different republics. The investigations on biosensors within the blank frame of the Research Program, "Newest Methods of Bioengineering," were reevaluated and moved in Russia under the supervision of the Ministry for Science, Advanced Education and Technology Policy of the Russian Federation, a successor of the State Committee for Science and Technology. The scientific Council for the field "biosensors" was organized and coordinates the work of Russian organizations.

The total volume of state financing of investigations on biosensors in 1992 in Russia was about 10 million rubles and is increasing. It should be stressed that the financing of scientific groups involved in the fulfilment of the State Program is carried out only by competition. One can add that private Russian capital as well as Western investments are beginning to become interested in biosensor research in Russia.

3. BIOSENSOR MARKETS

The creation of biosensors in Russia is aimed first at the needs of medicine (detection of low molecular mass compounds, proteins, creation of diagnostic systems for different diseases), control of processes in the food and fodder biotechnological industry, and the needs of ecology (detection of harmful substances under various conditions).

Despite early priorities in the fundamental development of this area (for instance, No. 316 discovery on the State Register of Discoveries—"Bioelectrocatalysis" had a priority date of 1978), only during the past few years has it become possible to organize in Russia the production of small quantities of biosensor devices. Recent progress in the field is due to the establishment of small and joint ventures at the leading research centers.

Some biosensor devices existing now in Russia are close in their characteristics to Western analogs. Systems for detection of metabolites and antibodies by semiconductor transducers and optico-electronic devices are at the stage of scientific and experimental elaboration in Russia. The coming 5-10 years will be characterized, apparently, by an increased interest in immunosensors for detection of different diseases, as well as DNA-based biosensors for detection of genetic diseases and factors destroying genetic material of living cells. However, it is still early to speak about the existence of a market value for biosensors in Russia because of their small volume (the evaluations of realized biosensor devices in the USSR at the end of 1990 showed that it has reached only 0.7-0.8 million rubles).

4. CURRENT DIRECTIONS

The papers of Russian scientists on creating biosensors are published in the journal, *Biotechnologiya*. In 1990 detailed reviews of these works were published in a special issue, "Biosensors," (series "Itogi nauki i tekhniki," Biotechnology series, Vol. 26, 1990). In 1992 in Moscow the first joint seminar on biosensors with scientists from Germany was organized. At different stages of realization there are joint projects on biosensors with researchers from Germany, Italy, China, and England.

In this volume, reviews written by Russian scientists are presented. The articles outline the main directions of investigations and developments in the field of biosensors within the framework of the Russian State Research Program "Newest Methods of Bioengineering." These reviews demonstrate the original approaches of different scientific groups in creating new biosensor types. On the whole, Western readers are not familiar with these works. We would hope that the acquaintance of the readers with these reviews will stimulate the contacts between scientists from different countries in the field of biosensors, a new, and quickly emerging area in biotechnology.

PRINCIPLES OF ELECTROCHEMICAL BIOSENSOR DEVELOPMENT

M. R. Tarasevich and V. A. Bogdanovskaya

OUTLINE

Advances in Biosensors
Volume 3, pages 5-29.
Copyright © 1995 by JAI Press Inc.
All rights of reproduction in any form reserved.
ISBN:1-55938-535-9

1. INTRODUCTION

In this review the results of investigations of electrochemical and enzymatic activity of glucose oxidase and glucose dehydrogenase on various electrode materials, the selective properties of polymeric membranes, and methods of preparation of the different enzyme microreactor types are presented.

The use of electrodes after special treatment or modification by various compounds makes it possible to decrease the potential of electrochemical oxidation of hydrogen peroxide. The use of enzyme reactors with supplementary membranes allows the selectivity to be increased and the range of determination concentrations of substrate to be extended for practical applications.

Further investigations of biosensors for glucose determination are directed towards the optimization of designed devices and a search for new sensitive elements for substrate determination.

In Japan, the United States, and a number of European countries various designs of glucose sensors have been developed, and a variety of instruments are in routine production [1]. Currently, priority has been given to gluconometers based on electrochemical principles.

An analysis of the literature shows that there still remains a steady interest in research in this field. Results that are obtained serve as a basis for the further improvement of the characteristics of the electrochemical gluconometer and its ever wider use in medicine and the food and pharmaceutical industries; this has allowed basic criteria to be worked out and recommendations concerning construction of biosensors for other substrates.

There is no routine production of any gluconometers in Russia and their importation from abroad is very limited because of economic reasons among which is the necessity of a constant supply of consumables: microreactors, electrodes etc. For the same reasons it is unlikely that the purchase of technology from abroad can be afforded.

In our opinion we are obliged and able to solve the problems of electrochemical gluconometer production by our own efforts with due account of specific features of medical service granted in Russia and the Russian market. Homemade instruments should meet the following requirements:

1. To be simple in design and reliable in operation.
2. Only homemade materials and accessories are to be used in the production.
3. Reasonable prices for the instrument and its spare parts.

It goes without saying that common requirements of sensitivity and selectivity of the gluconometer should remain within the generally accepted standards.

When we started our efforts to meet the above stated requirements, several versions of gluconometers had been described in the literature. Nevertheless, we had to cover all the stages in our search for an optimum solution. In this review we will dwell on the particulars of our study that might be interesting to our colleagues doing research in this field.

2. WAYS OF SELECTIVITY IMPROVEMENT

The biosensors in which the system of transformation of the substrate concentration quantity into electric signals is electrochemical belong to the class of electrochemical biosensors. The notion of a "biosensor" refers to a completely integrated system, that is to an analytical instrument. The sensitive element of the electrochemical biosensor is the enzyme electrode. The most difficult problem to solve in the course of building an enzyme electrode and a biosensor as a whole is coordination of electrochemical and enzymatic reactions while preserving the high specificity characteristic of the latter.

In order to increase the sensitivity and service life, the amount of immobilized enzyme per unit of the geometric surface should many times exceed the value which corresponds to a monolayer. The immobilized enzyme takes a certain volume restricted by a membrane or two and forms some sort of microreactor. That's why for the research, development, and optimization of a selective method of the substrate determination it is expedient to do analysis of micro- and macrokinetics of the processes of mass transfer occurring in the enzyme electrode.

Figure 1 shows a schematic view of the construction and variants of mass transfer organization correspondingly depicting the most important features of typical enzyme-electrochemical systems functioning.

The first of these systems describes an ideal case of immediate exchange of electrons between the enzyme active centre and the electrode. The second system deals with a mediator mechanism. As for the third one, it is devoted to the formation during the progress of the enzymatic reaction of an electrochemically active product. In the latter case, the total process consists of the following stages: diffusive or convective substrate (S) transfer to the membrane surface 1; insertion into and diffusion in a membrane with the constant γ_s of the transfer; enzymatic reaction of the substrate conversion into the product (P) with the Michaelis-Menten constant K_M in the microreactor; transfer of the product through membrane 2 with the constant γ^2_p and; electrochemical conversion of the substrate (or product) with the constant K_s of the velocity. The total current of the sensor ($j\Sigma$) can be described by the expression:

$$\frac{1}{j_\Sigma} = \frac{1}{j_{ed}} + \frac{1}{j_{id}} + \frac{1}{j_e} + \frac{1}{j_1} + \dots$$

These factors determine the sensitivity of the electrochemical biosensor. It follows that in every case it is required to do analysis of macrokinetic processes occurring in the enzyme electrode.

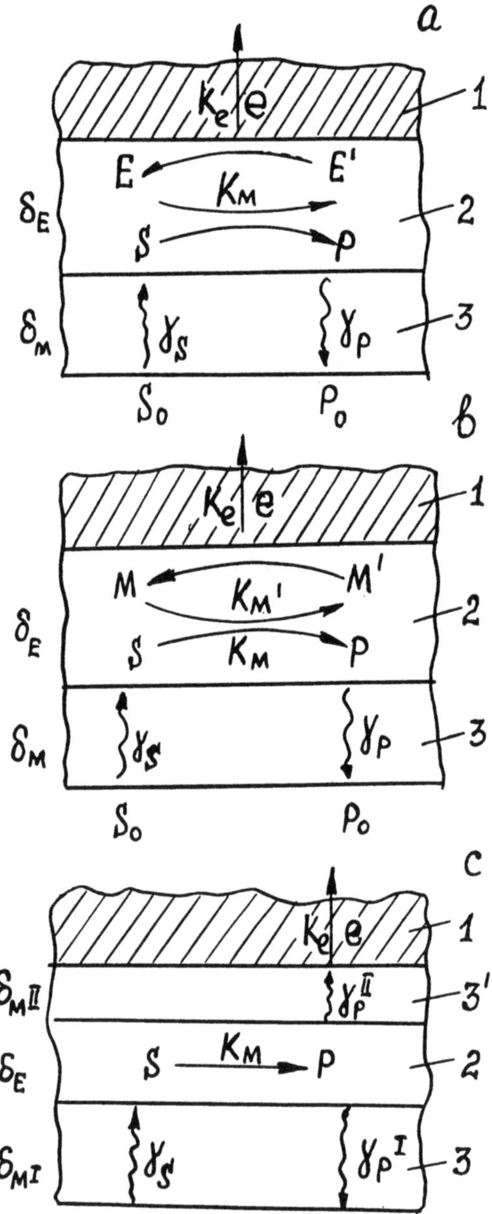

Figure 1. The scheme of construction and variants of mass transport organization for typical enzyme-electrochemical systems: (**a**) direct exchange of electrons between the enzyme active center and the electrode; (**b**) mediator mechanism of electron transfer between the enzyme active center and the electrode; (**c**) electrochemical transformation of the product of the enzymatic reaction. (**1**) electrode; (**2**) enzyme layer; (**3**) outer membrane; (**3´**) first membrane.

In order to find out velocities of the transfer in the membrane one has to take into account the distribution coefficient (K_S):

$$j_{id} \cong \gamma_s = D_s \cdot K_s \cdot /S_m$$

The velocity of the enzymatic reaction can be described by the Michaelis-Menten equation:

$$j_e = k_1 \cdot [E] \cdot [S]/K_M + [S]$$

The velocity of heterogeneous electron transfer is described by an equation of electrochemical kinetics:

$$j_1 \cong a \cdot K_1 \cdot \exp\{b \cdot \Delta E\}$$

E is the potential shift away from the value of equilibrium; a includes the quantity of the reagent surface concentration; the constant K_1 includes velocity of the adsorption on the electrode. It is seen that by changing the value ΔE one can vary within wide ranges of velocities of electron transfer.

The ratio between velocities at different stages depends on a number of conditions: the electrode construction, substrate concentration, terms of its supply, as well as membrane permeability, enzyme activity, the value of the electrode potential etc. A complete solution for this system is sought within the framework of analysis of a general macrokinetic equation on condition that the actual physical and chemical processes are properly described.

In view of the construction of enzyme electrodes we can be confined by the consideration of some extreme case. The most interesting of them seem to be the cases when it is possible to realize linear dependence of current quantity on the substrate concentration, namely, intra-and extra-diffusion condition for the substrate. Under these conditions it is necessary to solve the problem of selectivity. The main means to achieve this goal are as follows:

1. To make a proper choice of enzyme system.
2. To make use of membranes that will selectively transmit substrates and/or the product of enzymatic reactions.
3. To find the proper electrode material and/or method of electrode modification.
4. To determine correctly an optimum range of potentials and the method of measurement.

2.1 Choice of Enzymatic System

The properties of two enzymes: glucose oxidase (GOD) and glucose dehydrogenase (GDH) have been examined with a view to finding out the possibilities for their use in enzyme electrodes for glucose determination [2,3].

Glucose dehydrogenase is a cofactordependent enzyme performing the reaction:

$$\beta\text{D glucose} + \text{NAD} \xrightarrow{GDH} \text{D-gluconate} + \text{NADH}$$ according to which we can determine the glucose content for instance, measuring the current of oxidation of the enzymatic reaction product—nicotinamide adenindinucleotide (NADH).

GDH was adsorbed on carbon material with the activity of the enzyme retained judged by the NADH oxidation current when introducing glucose into the solution containing NAD. In the absence in the solution of the cofactor, NAD, no dependence of the oxidation current at the potential +0.2 V (Ag/AgCl) on the glucose concentration was observed. The Redox potential of the transfer NAD$\rightleftharpoons$NADH is −0.55 V and NADH oxidation on carbon materials at potentials more positive than +0.2 V proceeds with an overvoltage.

In Figure 2 the dependence of the NADH oxidation current on the glucose concentration is presented. In order to reduce the overvoltage of NADH oxidation we used the mediator N, N'-dimethyl-7-amino-1, 2-benzophenoxazinyl, kindly provided to us by Professor G. Johansson (Lund, Sweden).

In the presence of the mediator (M) the following reactions are taking place at the electrode and in the solution:

$$\text{NADH} + \text{M}^+ \rightarrow \text{NAD} + \text{MH}$$
$$\text{MH} \rightarrow \text{M}^+ + \text{H}^+ + 2e$$

The anodic reaction proceeds at a potential close to zero when minimum oxidation or reduction of the other components of body fluids occurs; thanks to that, high selectivity of the glucose determination is ensured. Another important advantage of enzyme electrodes with GDH is that the reaction is independent of the oxygen content in the assay.

However the use of enzyme electrodes based on GDH for the construction of real biosensors entails some difficulties since a large consumption of the cofactor, NAD, and the mediator are required. The cofactor and mediator can be adsorbed on the electrode but the adsorption is not stable enough and the amount of the adsorbed NAD is insufficient for taking measurements in a wide range of concentrations. In Figure 2 (curve 3) dependence of oxidation current on glucose concentration is shown when GDH, NAD and the mediator are adsorbed on a carbon electrode.

It should also be noted that the active center of the adsorbed GDH, does not participate in the electron transfer between the electrode and the substrate, in voltammograms there are no maximums characterizing redox transformations of the enzyme active centre i.e., realization of conditions for direct bioelectrocatalysis.

The second enzyme under investigation, glucose oxidase, belongs to the FAD (flavinadenindinucleotide)-containing group of enzymes and accelerates the reaction of glucose oxidation with oxygen or another electron acceptor:

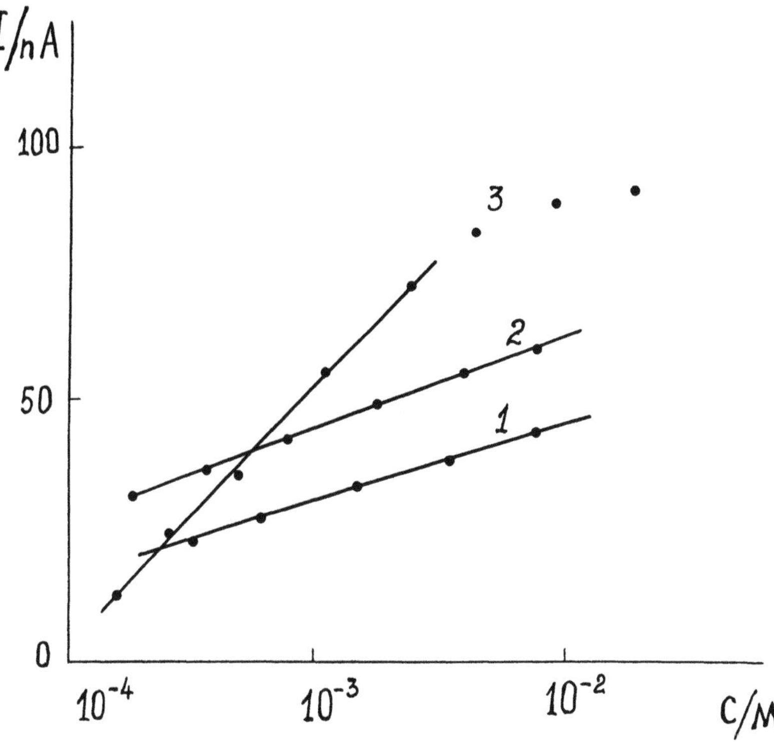

Figure 2. The dependence of NADH oxidation current on the glucose concentration at a carbon electrode with adsorbed GDH (C_{NAD} = 2 mM) at potential, V: 1 - 0,3; 2 - 0,1. (Curve 3) the same dependence on carbon electrode with adsorbed GDH, NAD, and mediator (benzophenoxazinyl) at E = 0.0 V. Phosphate buffer solution pH 7.

$$\beta\text{-D-glucose} + O_2 \xrightarrow{\ GOD\ } \text{gluconic acid} + H_2O_2$$

In this connection it is possible to do determine glucose with the help of GOD by the decrease of the current value of the electron acceptor reduction (O_2, quinones, ferrocenes, etc.) or oxidation of the enzymatic reaction product (H_2O_2, hydroquinone, etc.). Particularly promising is realization of direct bioelectrocatalysis for the immobilized GOD because, in this case, we can obtain an electrode being exclusively selective toward glucose.

One of the important stages for the realization of bioelectrocatalysis is the presence of exchange of electrons between the electrode and the enzyme-active center. We demonstrated [2,4] that GOD immobilized on carbon materials takes part in redox transformations.

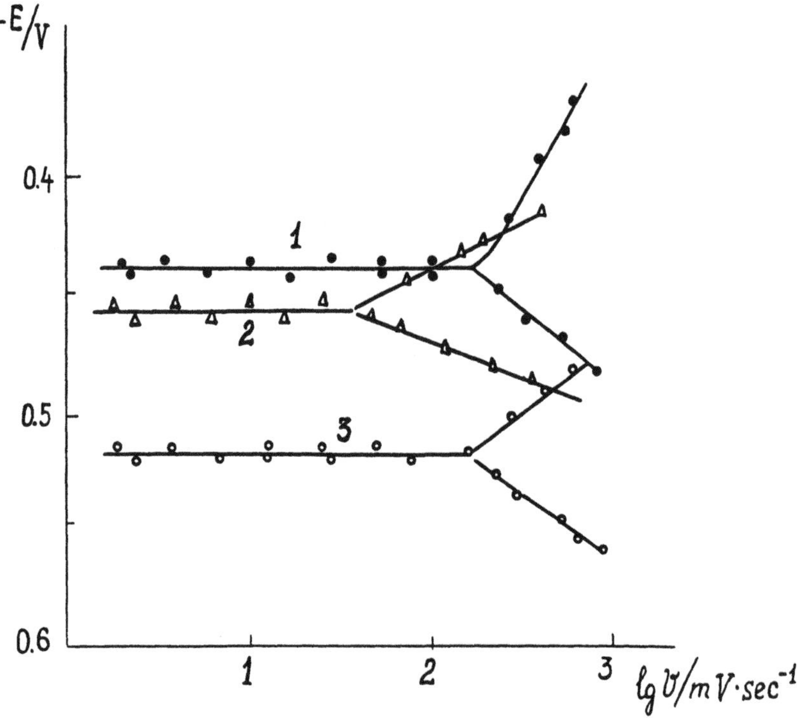

Figure 3. The dependence of the maximum potential positions on the potential sweep rates for enzymes: (**1**) GOD (**3**)   PhHD, and (**2**) FAD, immobilized on carbon materials.

Based on the dependence of the maximum potentials position (Figure 3) on the potential sweep rates for the enzyme and free FAD, as well as the comparison with kinetic parameters (Table 1) of redox transformation of another FAD-containing enzyme, phenolhydroxylase (PhHD) [5], we came to the conclusion that the FAD included in the glucose oxidase active center participates in redox transformations.

First of all independence of the maximum potential position upon potential sweep rates is maintained for the enzymes up to higher velocities compared with free FAD. Besides, the potential displacement of GOD and PhHD redox transformations is 0.2 V while FAD experiences redox transformations with the potential. The values of slopes $\partial E/\partial\lg v$ (Table 1) are also essentially different. All those differences drive us to the conclusion about the exchange of electrons between the electrode and GOD active center.

Nevertheless, only the electrode with immobilized GOD reveals some catalytic activity in the reaction of glucose oxidation. Enzyme activity was determined

Table 1. Redox Potentials and Kinetic Parameters of Redox Transformations of FAD, PhHD, and GOD (pH7)

Compound	$E_{red/ox}$ in solution, V	$E_{red/ox}$ on electrode, V	v, Vs^{-1}	$\partial E/\partial lgv$, V	$\partial E/\partial pH$, V
FAD	0.44	0.46	<0.025	0.0	0.06
			>0.025	0.035	—
PhHD	0.43	0.46	<0.25	0.0	0.06
			>0.25	0.06	—
GOD	0.30	0.43	<0.1	0.0	0.06
			>0.1	0.20	—

from the oxidation current of hydrogen peroxide formed as a result of the enzymatic reaction. A possible supposition [6-8] put forward for the reason why direct bioelectrocatalysis is unavailable could be retarding of the oxidation stage of the enzyme active center.

In order to accelerate electron transfer previous workers modified the enzyme molecule [6] or the electrode surface with ferrocene [7,8]. In the latter case [8], while modifying the surface with polycationic redox polymers, the current of glucose electrooxidation was obtained at potentials more positive than +0.2 V (Ag/AgCl). It seems to us that another explanation of the absence of glucose oxidation biocatalytic current on the nonmodified surface is likely: being adsorbed on carbon materials the active center is so oriented that the exchange of electrons with the electrode is feasible but the active center is inaccessible to glucose, while in the presence of a polycationic polymer a GOD orientation is realized that provides for electron exchange between the electrode, the active center and the substrate. However, these enzyme electrodes have not found actual application because of difficulties in production and the nonreproducibility of the results.

Analysis of data available in the literature and of the results of our own studies shows that the most promising way of constructing real electrochemical biosensors is by the use of glucose oxidase with the recording of the current due to hydrogen peroxide oxidation. Based on this principle glucose detection achieves high sensitivity and provides for the possibility of glucose determination over a wide range of concentrations.

The main disadvantage of the hydrogen peroxide amperometric detector is its low selectivity. Generally speaking, the selectivity of the glucose determination depends on the enzyme and electrochemical specificity of the detector. GOD is an extremely selective enzyme but at the potential of hydrogen peroxide oxidation numerous compounds present in body fluids can be oxidized on the electrode, such as ascorbic acid, uric acid, free amino acids, glutathione, medicinal preparations (like paracetamol), and others. With this in mind we aimed our research at studying physical and chemical characteristics of both individual components and a whole sensitive element, and working out recommendations for the creation of a selective bioelectrochemical sensor of the amperometric type using the enzyme, glucose oxidase.

2.2 Utilization of Membrane Selectivity
Transmitting Substrate or Enzymatic Reaction Product

At first sight the simplest way providing for selectivity of glucose determination is the use of a selective membrane.

A schematic construction of an enzyme reactor is shown on Figure 4. The inner membrane of acetylated cellulose is a cross-linked polymer and provides for preferential passage of H_2O_2 to the anode due to the smaller size of a peroxide molecule as compared with the other substrates. It has been shown that the permeability to interferents of the cellulose-based membrane is the lower, the higher the degree of cellulose acetylation. However, in this case the selectivity increase also results in additional diffusive obstacles to the diffusion of H_2O_2.

In order to ensure selectivity of the external layer (Figure 4) it is necessary to develop a membrane permeable only to glucose and oxygen. Our research was aimed at the study of nuclear lavsan filters, today more often termed "tracking membranes." Advantages of tracking membranes are as follows: narrow scattering of the radius of pores, mechanical strength, thermal and chemical stability, and the possibility of chemical modification aimed at imparting to the surface previously selected properties.

In Table 2 we compare diffusion penetration factors of nuclear filters for glucose and ascorbic acid depending on the size of pores and charge of their surface.

Comparison of integrated factors of the diffusion penetration of nuclear filter samples with volumetric efficiency of the glucose and ascorbic acid diffusion equal to $5.2 \cdot 10^{-6}$ cm^2/s and 10^{-5} cm^2/s, accordingly, shows that the values of diffusion factors in the pores are smaller than in the volume of the liquid. The diffusion penetration factor for ascorbic acid decreased to a greater degree. This is

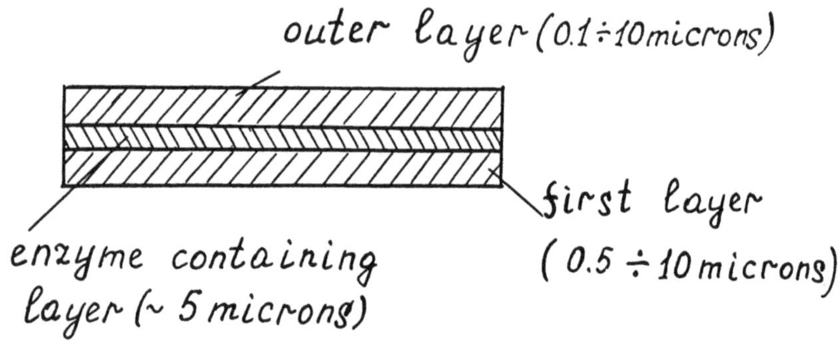

Figure 4.   A schematic construction of the enzyme reactor.

Table 2. Structural, Electrokinetic, and Selective Characteristics of
Nuclear Lavsan Filters (Polyethyleneterephthalate)

Diameter of pores, Å	Pores Density, Number of pores per cm^2	S_{pores}, $10^{-3} cm^2$	ξ, Density of Charge, $V\ 10^{-4} Cm^{-2}$	P_{spec}*, $10^{-7} cm^2 s^{-1}$ glucose	ascorbic acid
143	5	8	0.30 −2.18	1.1	0
225	2	8	0.043 −3.1	1.1	0
314	5	39	0.020 −1.46	4	4.6
415	2	27	0.021 −1.54	1.1	7

Notes: *Specific diffusion penetration factor was calculated by the expression: $P_{spec.}$ - P/S, where

$$P = \frac{K_{cell}}{C_{initial}} \frac{\Delta C}{\Delta t}; \ S = \text{the total area of pores, } K_{cell} \text{ is the constant of cell.}$$

connected with the fact that molecules of ascorbic acid at neutral pH values are in a dissociated form $K_1 = 7.9 \cdot 10^{-5}$, $K_2 = 1.6 \cdot 10^{-12}$ and the pore walls of the filters are negatively charged. The data cited shows that from the selectivity point of view the optimum membranes are those with pore sizes of less than 22Å.

Since tracking membranes with a pore diameter smaller than 225 Å have less mechanical strength, we utilized membranes with pore diameter of 465 Å and subjected them to silanization.

In Table 3 test data of the silanizied membranes are given. From the data in Table 3 it follows that with the pore diameter decrease and the increase of the pore wall's negative charge, the permeability for ascorbic acid is reduced. It should be noted that this reduction for ascorbic acid cannot be explained only by electrostatic interactions. The selectivity effect might be stimulated by the fact that ascorbic acid is a small addition to the background solution (all the measurements were made in 0.15 M NaCl) where Cl−anions are competitive relative to the ascorbic acid anions. By virtue of the fact that the mobility of chlorine ions is higher than that of the anions of ascorbic acid, the flow of negative charge to balance the flow of positive charge is mainly due to the transfer of chlorine anions. This leads to the selectivity relative to the ascorbic acid anion. A similar explanation has lately been

Table 3. Physicochemical Parameters of Silanized Membranes

Type of Treatment	Diameter of pores, Å	ξ, V	Density of Charge, $10^{-4} C\ cm^2$	$P_{spec.}$, $10^{-7} cm^2 s^{-1}$ glucose	ascorbic acid
Without Treatment	465	0.013	−9.3	22	3.1
Silanizataion with Dimethylchlorine silan in Vapors	420	0.016	−12	9	1.0
Silanization with Dimethylchlorine silan in Vapors	260	0.030	−22	16	0

successfully applied to the processes of reverse-osmotic separation of multicomponent electrolyte solutions [9]. Thus, utilization of modified tracking membranes made of polyethylene terephthalate allows ascorbic acid penetration to the surface of biosensor indicator electrodes to be eliminated.

2.3   Selection of Electrode Material and Methods of Electrode Modification

When selective membranes are unavailable components of body fluids penetrate to the surface of a platinum indicator electrode and are able to oxidize at a potential of +0.6 V, the optimum for oxidation of hydrogen peroxide on this electrode. Some polarized curves of electrooxidation of several components available in blood are shown in Figure 5.

It follows that practically all the components studied are oxidized on the indicator electrode with this making a contribution to the measured magnitude of current and accordingly to the determined quantity of the glucose content. Utili-

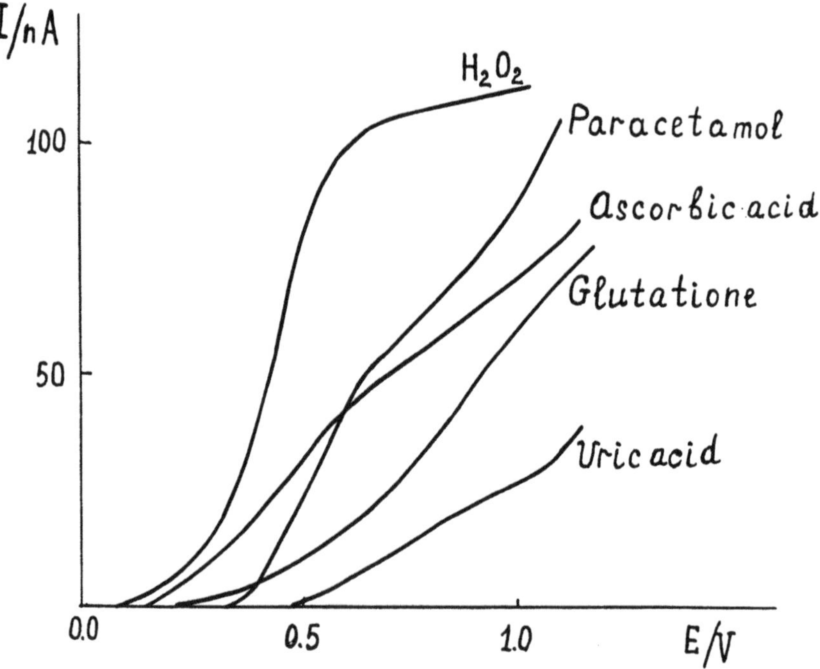

Figure 5. The polarization curves of electrooxidation of different substrates on a platinum electrode. pH 7, 0,9% NaCl solution.

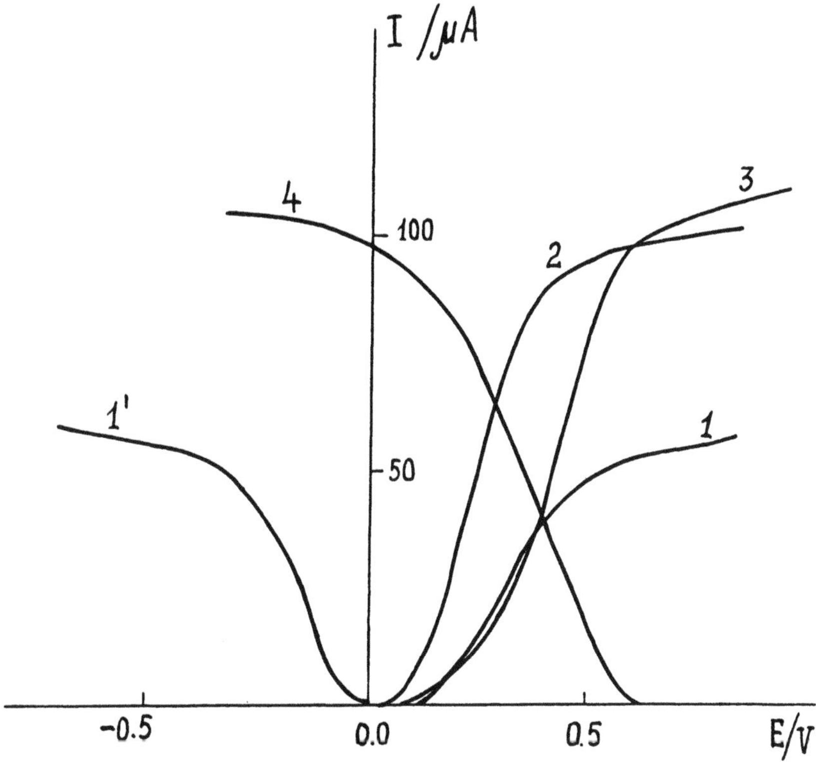

Figure 6. The polarization curves of electrooxidation (1-3) and electroreduction (1',4) of H_2O_2 on different electrode materials: (1,1') carbon black modified by pyropolymer of tetramethylphenylporphyrine Co; (2) alloy Au-Pd; (3) Pt; (4) carbon black with immobilized peroxidase.

zation of other electrode materials allows for reduction of the potential required for hydrogen peroxide conversion.

The polarization curves typical of hydrogen peroxide conversion are given in Figure 6. You can achieve a certain displacement of the hydrogen peroxide oxidation potential by using an alloy of palladium with gold [10] or modifying the surface of carbon materials with pyrolized porphyrin complexes of cobalt or iron (Figure 6) [11]. The highest effect in reducing the potential of the hydrogen peroxide transformation was achieved by modifying the surface of the carbon material with peroxidase (POD) [12]. POD immobilized on carbon participates in redox transformations and retains its activity. Hydrogen peroxide electroreduction under conditions of direct bioelectrocatalysis takes place within the range of potentials +0.6 V to 0.0 V. This gives the possibility to record hydrogen peroxide

Table 4. Influences of the Electrode Potential on Michaelis Constant

E, V	0,5	0,4	0,2	0,2
$K_M \times 10^4$	2.6	4.0	7.2	16

at a potential close to zero where body fluid components are practically neither oxidized nor reduced. A study of POD adsorption on carbon materials and the kinetics of H_2O_2 electroreduction showed that the most probable rate limiting step was the enzymatic reaction. Dependence of the reaction velocity on H_2O_2 concentration obeyed the Michaelis–Menten equation.

In Table 4, K_m values at different potentials of H_2O_2 reduction are presented. The bigger the potential displacement the higher the constant of velocity. At higher concentrations the maximum velocity of immobilized enzyme was observed at more negative potentials (Figure 7). This allows the determination of

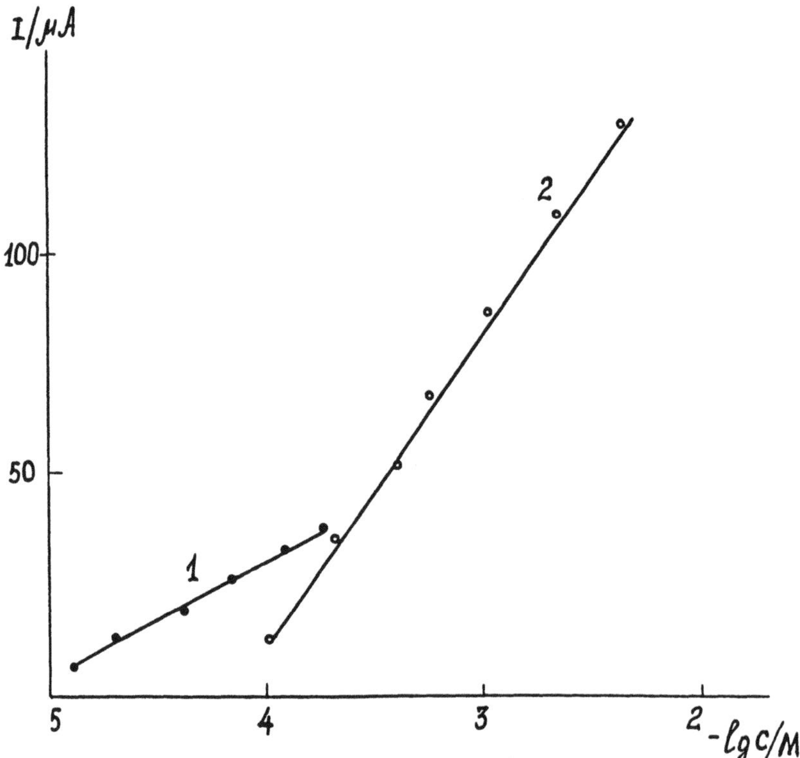

Figure 7. The dependence of the current due to electroreduction of H_2O_2 on the glucose concentration at an electrode with peroxidase adsorbed on carbon black at a potential, V: (1) 0.60; (2) 0.45. pH 7, phosphate buffer solution.

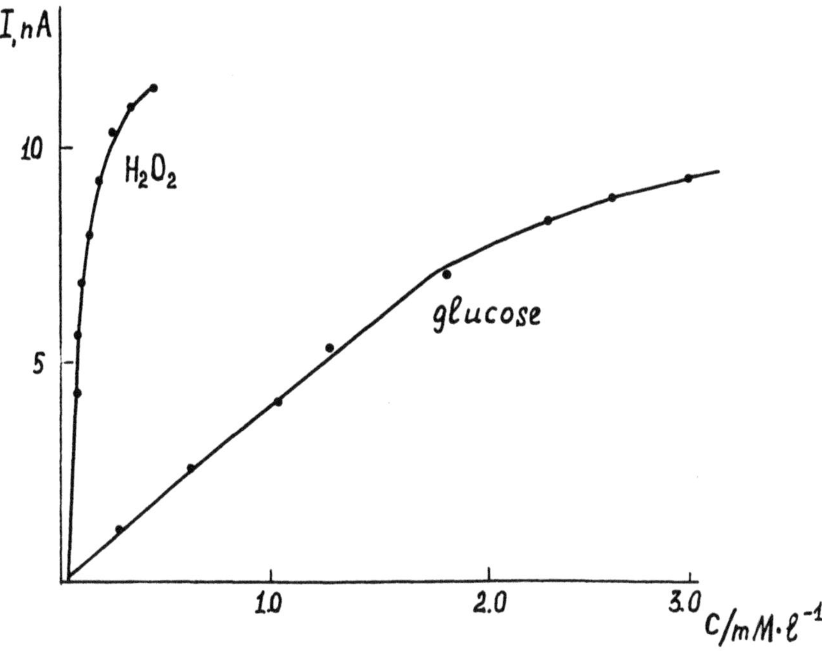

Figure 8. The dependence of the reduction current on H_2O_2 (1) and glucose (2) concentration obtained with a pyrographite electrode with coadsorbed GOD and POD at $E = 0.01$ V vs. SCE.

H_2O_2 and accordingly glucose concentrations within the range of $5 \cdot 10^{-4}$ to $5 \cdot 10^{-2}$ M, while maintaining an electrode potential of less than +0,2 V. For the concentration determination in the range of $5 \cdot 10^{-5}$ to $5 \cdot 10^{-4}$ M $E = 0.6$ V is considered to be the optimum value.

To prevent desorption, POD can be immobilized into the polymer film. By using pyrrole under conditions of electrochemical polarization, polymeric enzyme electrodes were synthesized on carbon material [13].

Calibration curves for H_2O_2 and glucose are presented in Figure 8. The measurements were made with the potential close to zero which provided for the selective determination of glucose.

2.4 Mediators and Method of Measurement

As it was shown above, changing the nature of the electrode material or modifying the surface of the electrode can reduce the potential at which the electrochemically active component is recorded.

Table 5. Redox Potentials and Potential of Oxidation of Ferrocene and its Derivatives on Carbon Electrodes

Ferrocenes	$E_{red/ox}$ V	E_{ox} V
Fc(CH$_3$)$_2$	0.241	0.280
Fc	0.310	0.365
FcCOOH	0.275	0.399
Fc(COOH)$_2$	0.395	0.460

Another route to potential reduction is the use of mediator couples with the mediator's oxidized form being an electron acceptor in the glucose oxidation reaction. The couple of quinone-hydroquinone is widely used for these purposes. Quinone oxidation on carbon materials occurs with $E = 0.3$ V. However, the couple is unstable with the formation of radical particles and polymerization processes going on resulting in the electrode activity decreasing.

Ferrocenes have higher stability. Among the ferrocene derivatives studied by us electrodes modified by dimethylferrocene have stable characteristics and low potential of oxidation.

In Table 5 values of the oxidation potential of ferrocene derivatives investigated are cited.

Thus the mediator couples allow the selectivity of the determination to be increased but they are not stable enough and complicate the system. The most promising area of application of electrodes of this type is in disposable systems [14].

Certain progress in discrimination of the selective signal proportional to the glucose concentration can be achieved by measuring the current typical for the half-wave potential rather than the limiting current of the hydrogen peroxide oxidation. Kinetics of the oxidation of various substrates differs and thanks to this it is possible to avoid oxidation of interferents. Moreover, the time of measuring in this case is reduced providing for a longer service life of the sensitive element.

3. SYSTEMIC SERIES OF GLUCONOMETERS

The above data on the study of development of sensitive and selective electro-chemical elements gave us a possibility to work out in our group a systemic series of gluconometers based on GOD. We believe these instruments are the most suited to the conditions of medical service in Russia.

Glucose analyzers are intended for operation under different conditions and according to this they have different characteristics presented in Table 6. All the gluconometers operate on a two-electrode circuit. In most cases we use the method of amperometric signal detection. For disposable elements we have also developed a version providing for the integration of the quantity of electricity proportional to the glucose content in the blood test.

Table 6. Gluconometer Characteristics

Purpose of Instument	C, nM	Number of Measurements Per Hour	τ90% s	Life Time, h	Error, %
Physician's room	0.0-20.0	30	50	600	7-8
System "Biostator" (Flow-Through)	0.0-20.0	continuous	–	50	5
Personal with Multiple Sensitive Element	2.0-15.0	20	120	200	8-10
Personal with Disposable Sensitive Element	2.0-15.0	10	180	–	10

3.1 Gluconometer for Physician's Room

This instrument (Figure 9) is the basis for a whole series of gluconometers. The electronic circuit of the instrument features simplicity of configuration, it is based on the analog-digital principle and has stable operation in the range of 1 to 100 nA that ensures measurements in the glucose concentration interval with the scale of the instrument indicating the value corresponding to the glucose concentration in blood of 2 to 20 mM (after the dilution - 0.2 to 2.0 mM). The instrument provides for the possibility of periodic correction of sensitivity with the help of standard glucose solutions. The instrument is designed for use by medical personel experi-

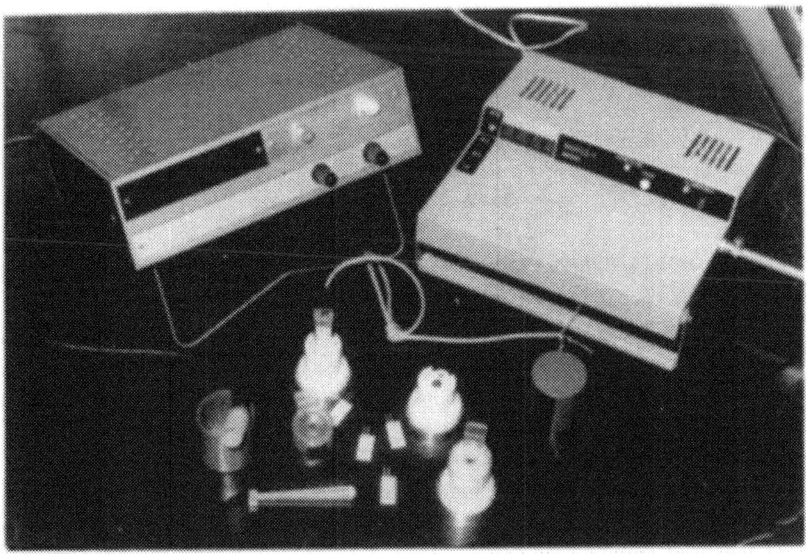

Figure 9. The gluconometers and different type of electrochemical elements.

enced in tests on blood samples and is complete with simple devices for exact blood dilution and calibration with a standard glucose solution. Samples of 20-25 μl of blood are required for the analysis.

The electrochemical unit of the instrument includes a platinum indicator electrode and Ag/AgCl counter electrode. The ratio of the electrode areas is matched to ensure the performance of an electrochemical element without regeneration for a year. The measured current of H_2O_2 oxidation is recorded at certain intervals providing for the 90% value of response. The membrane designed for taking 500-600 measurements with the selectivity maintained has been developed on the basis of the research to study properties of the membranes, enzyme and electrochemical unit. Figure 4 shows the construction of an enzyme reactor where the external membrane is permeable to glucose and oxygen and stable to fouling with blood proteins. The internal membrane (the cellulose cross-linked polymer or its derivatives) practically is only permeable to H_2O_2 and this creates additional difficulties for the penetration to the indicator electrode of blood components interfering with the glucose determination. The characteristics of these membranes and the whole sensitive element are cited in Table 6. The new generation of membranes ensures more stable operation of the reactor and thanks to a special ring the membrane replacement is facilitated.

Use of periodic calibration allows to a considerable extent for elimination of the undesirable effect of temperature on the measured quantity of the hydrogen peroxide. Within the temperature range of 15-45° C this value is 1-1.5 nA/°C. In practice the temperature influence is leveled thanks to the calibration made after every 15-20 measurements or an hour of operation, and subsequent measurements are made in the same temperature mode as the calibration. It is proposed to introduce temperature compensation into further modifications of the instrument.

At present the instrument is being tested in a number of clinics in Russia, Yugoslavia, China, and other countries.

3.2   Flow-Through Analyzer

The flow-through version of an electrochemical analyzer was developed for the "Biostator" system used in departments of acute or postoperative therapy.

The interior of the electrochemical unit is shown in Figure 10. The construction of the electrodes and microreactor do not differ much from the one described above; the most important problem is maintenance of the hydrodynamic mode of the flow and constant temperature. To date the service life of the running-through sensitive element attained is about 50 hours (Figure 11).

The "Biostator" system has a special device for sampling microquantities of blood and its dilution. The digits on the scale (Figure 11a) correspond to the flow rate of the fluid analyzed and the digits below conform with the time of interruption of the measurements. In Figure 11b dynamic characteristics of the sensor corresponding to a change of the glucose concentration in blood are presented.

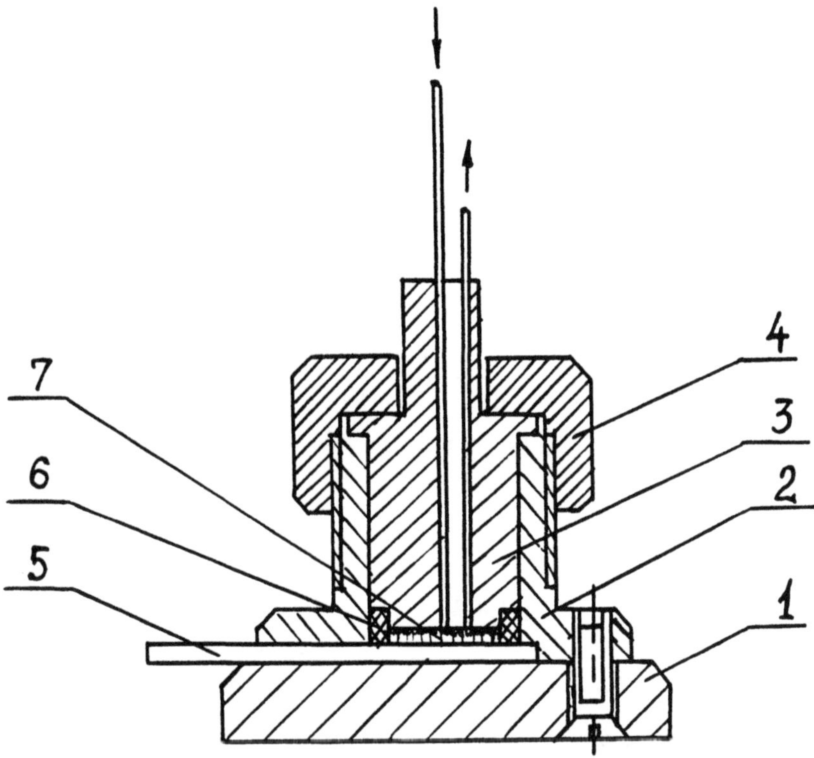

Figure 10. The electrochemical flow cell for flow-through analyzer: (**1**) base; (**2**) body; (**3**) insert; (**4**) screw-nut; (**5**) electrode; (**6**) enzyme; (**7**) outer membrane.

3.3 Personal Analyzers

Two types of sensitive element for personal analyzers have been developed simultaneously: multiple use and disposable ones.

The electrode unit of the sensor for multiple use is made on a ceramic base. The process flow diagram and sequence of operations are presented in Figure 12a. Initially with the help of two photopatterns the layers containing platinum and silver are applied on a ceramic plate by the group method (operation 1,2). Then a special heat treatment is applied with the consequent removal of organic binders from the paste and partial inclusion of the metal into the ceramic base (3). This ensures high adhesion of the electrodes to the base and creates preconditions for the multiple utilization of the electrode unit. The average electrode thickness is about 100Å. Then with the help of a special photopattern some insulating paste is applied

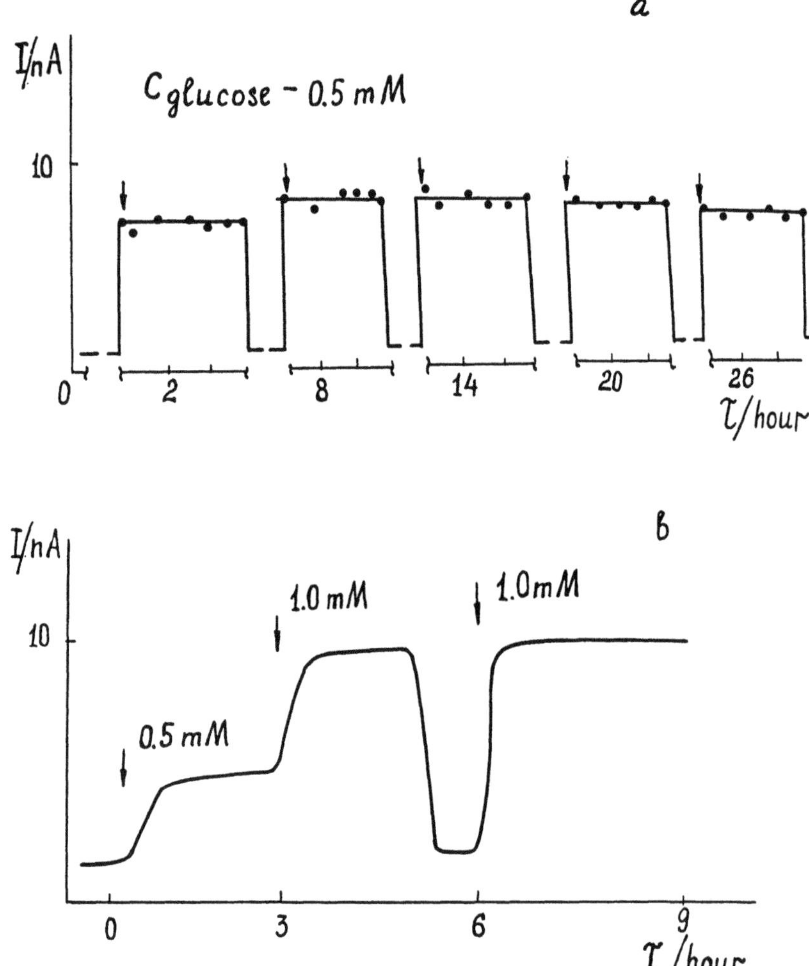

Figure 11. Characteristics of the sensitive elements for flow-through analyzer. The time between measurements is 18 hours. (**a**) dependence of the measurement current on the time; (**b**) dynamic characteristics of the sensor.

on the whole surface of the ceramic base (4). By subsequent heat treatment it is fixed to the surface (5).

After the plate is cut into separate electrode units, electrochemical electrode treatment is required (6,7). This first includes chlorination of the silver electrode and production of the Ag/AgCl-electrode. Second, electrochemical activation of

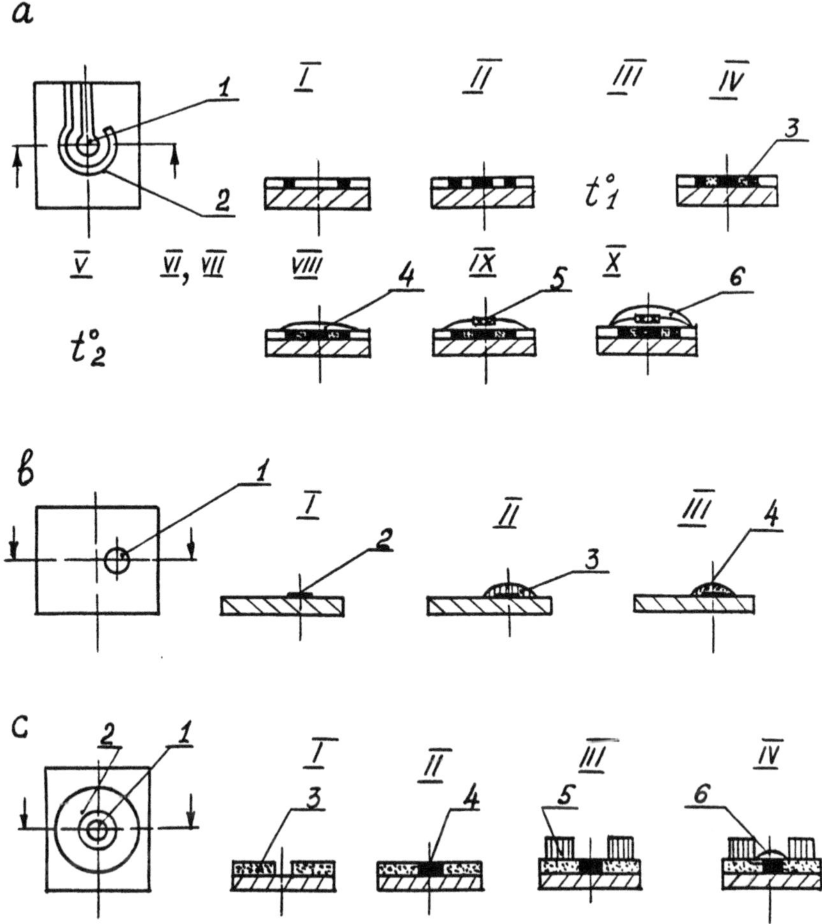

Figure 12. The diagram and sequence of operations for preparing different types of sensitive elements. Explanation in text.

the platinum electrode is required for imparting to it high electrochemical activity relative to the hydrogen peroxide electrooxidation reaction.

In Figure 13 curves of the platinum electrode charging before and after activation are given. After the electrochemical preparation the roughness factor of the platinum electrode is about 10-15.

In the course of further operations an enzyme microreactor itself is formed (operations 8-10). First the internal cellulose acetate membrane is applied on the electrode unit. This membrane plays the main role in providing selectivity while taking

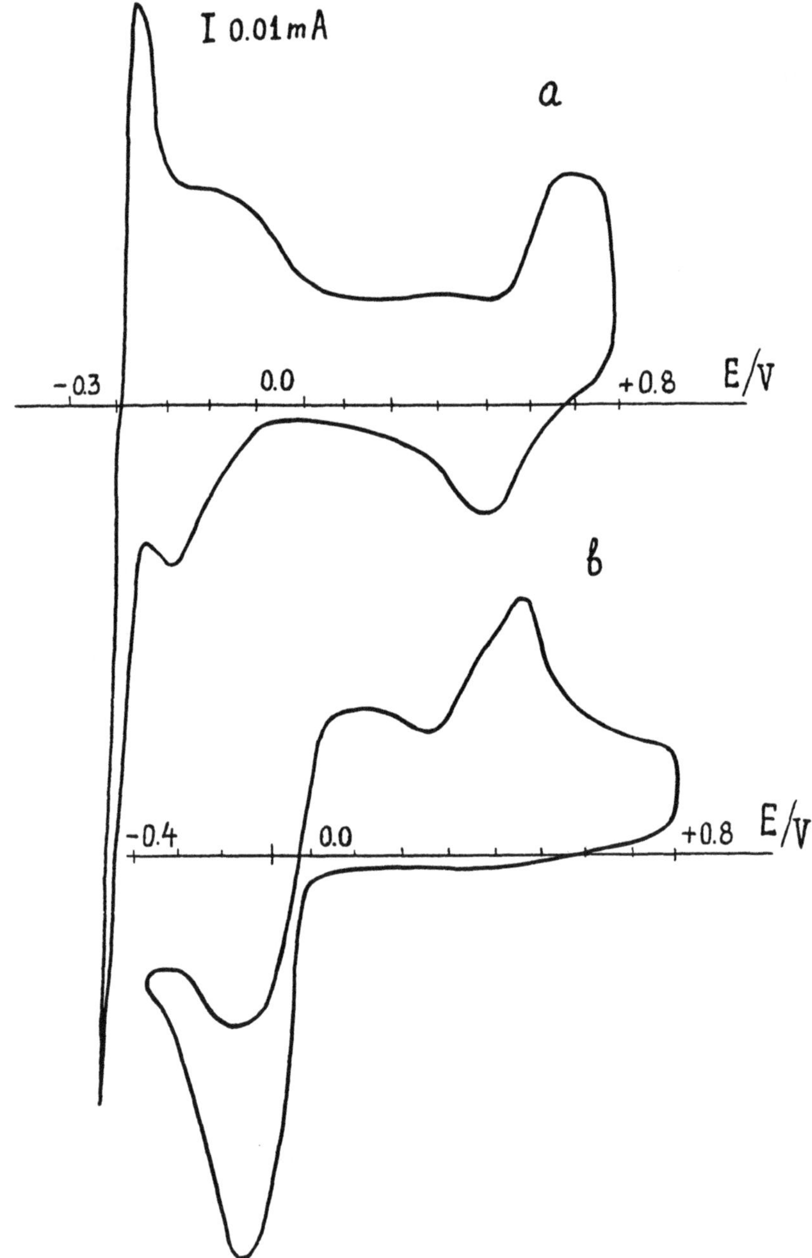

Figure 13. Cyclic voltammograms at a platinum-paste electrode: (**a**) after electrochemical treatment; (**b**) before electrochemical treatment. $v = 0.05$ V s^{-1}, 0.1 N H$_2$SO$_4$.

measurements with nondiluted blood. Then with the help of a special device GOD in polymer in dosed and distributed only on the surface of the platinum electrode. After completion of the polymerization process the external membrane is applied.

Built by this method, electrode units are reproducible in their characteristics and ensure sufficiently stable operation for at least 2 weeks or about 50 measurements without additional calibration.

For the production of disposable individual elements, electroconductive carbon paper was used as a backing. In a simpler version (Figure 12b) of the disposable element there is only an operating electrode with the enzyme reactor. The counter-electrode (Ag/AgCl) is permanently built into the contact device of the measuring instrument (details of its construction will not be discussed here). The technology

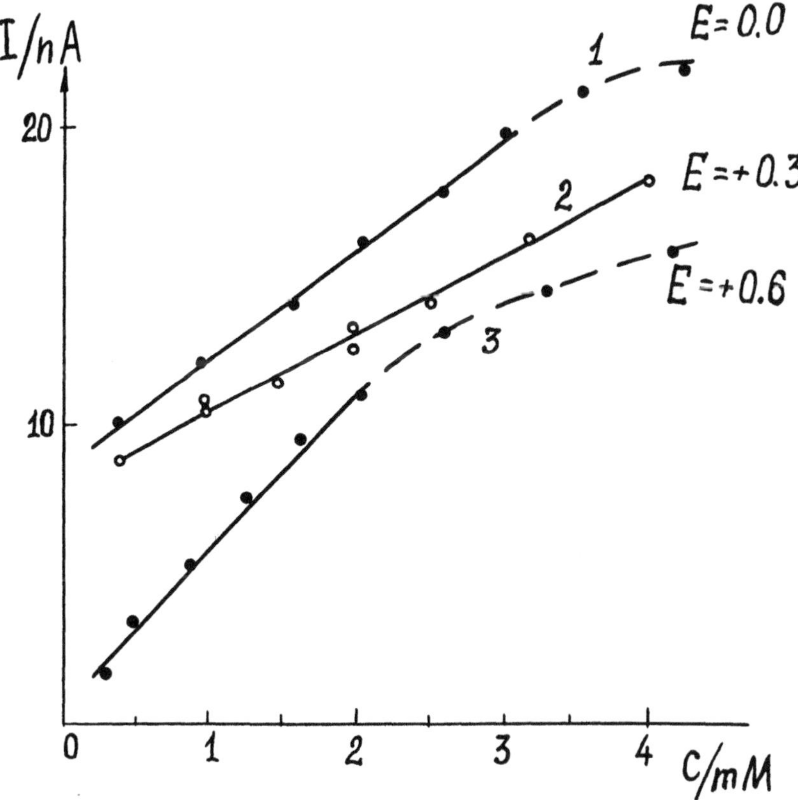

Figure 14. Response of biosensors versus the concentration of the glucose with different types of sensitive elements: (**1**) working electrode with adsorbed POD; (**2**) working electrode modified by dimethylferrocene; (**3**) Pt electrode.

of the operating electrode formation is based on using as‧ a modifier hydro-quinone, dimethylferrocene, or peroxidase. Modifying compounds were, first adsorbed on the highly dispersed carbon carrier carbon black. Then the modified carbon black was applied on the carbon paper (1) with the help of a binder. Following this step, the cellulose acetate membrane was formed on the electrode (2) and GOD was immobilized.

Figure 14 shows dependence of the current at this electrode on the glucose concentration. Use of the couple, quinone-hydroquinone, permits determination at E 0.3 V. Since carbon materials have low activity in electrooxidation reactions of organic substances, this potential provides for satisfactory selectivity. When dimethylferrocene was used, the potential +0.28 V was maintained on the electrode. A disadvantage in the practical application of such individual elements is the necessity of their preliminary reduction immediately before the application.

This disadvantage can be overcome by way of using carbon black modified with peroxidase. Direct bioelectrocatalysis of the hydrogen peroxide in the range of potentials of 0.0-0.2 V is ensured on this electrode. Such electrodes are advan-tageous in the area of low glucose concentrations where a contribution of the background current is relatively high (Figure 14).

A diagram of a more complicated electrode unit is shown in Figure 12c. By means of photopatterns, a double-electrode element is formed on electrocon-ductive paper. Initially the insulating polymeric interlayer is applied restricting the area of the operating electrode. Then the operating electrode, made of carbon black, is applied with a micropromoted plate (2). On the insulating interlayer we applied the counter-electrode on the base of the oxide with the appropriate redox-potential (3). The enzyme microreactor of either construction is formed on the operating electrode. Figure 14 shows the dependence of current on the glucose concentration at disposable electrodes of this type.

CONCLUSION

In this review we have presented the results of our research in the field of selective electrochemical biosensor construction as well as data on sensors intended for practical application, the outcome comprises scientific recommendations and possibilities for production.

Today the market for biosensors in Russia is practically nonexistent. This fact puts before researchers a conflicting problem. On the one hand, it is necessary to quickly work out some scientifically grounded prerequisites for the development of a series of biosensors for various applications. On the other hand it is expedient to conduct basic and exploratory studies of new ways to provide for high sensi-tivity and selectivity for the construction of electrochemical biosensors for future generations. We suppose that at the present time, before we move forward, we should start output of biosensors of the first generation acceptable from the producer and consumer points of view as far as the production conditions, characteristics, and as prices are concerned.

REFERENCES

[1] Turner, A.P.F., Karube, I., Wilson, G.S. (Eds.) (1987). *Biosensors: Fundamentals and Applications.* Oxford University Press, Oxford, p. 546.

[2] Atanasov, P., Bogdanovskaya, V.A., Iliev, I., Tarasevich, M.R. et al. (1989). Redox reactions of glucose oxidase on carbon materials. *Electrochem.* (Russia), **25**, 1480-1486.

[3] Bogdanovskaya, V.A., Iohansson, G., Tarasevich, M.R. (1989). Using dehydrogenase for analytical aim. *Chemical Sensors-89.* (Leningrad), **3**, 219.

[4] Bogdanovskaya, V.A., Tarasevich, M.R., Hintsche, R., Scheller, F. (1988). Electrochemical transformations of proteins adsorbed at carbon electrodes. *J.Electroanal.Chem.*, **253**, 581-584.

[5] Bogdanovskaya, V.A., Fridman, V.A., Tarasevich, M.R. (1992). Electrochemical reactions of phenolehydroxilase immobilized at pyrographite electrode. *Electrochem.* (Russia), **28**, 32-38.

[6] Degani, Y. Heller, A. (1987). Direct electrical communication between chemically modified enzymes and metal electrodes. 1. Electron transfer from glucose oxidase to metal electrodes via electron relays, bound covalently to the enzyme. *J. Phys.Chem.*, **91**, 1285-1289.

[7] Ianniello, R.M, Lindsay, T.J., Yacynych, A.M. (1982). Differential pulse voltametric study of direct electron transfer in glucose oxidase chemically modified graphite electrodes. *Anal.Chem.*, **54**, 1098-1101.

[8] Wrighton, M.S. (1986). Surface functionalization of electrodes with molecular reagents. *Science*, **231**, 32-37.

[9] Yaroschuk, A.E. (1989). Separation of ions on charge membranes. *Chem. and Technol. Water,* **11**, 867-883.

[10] Gorton, L. (1985). A carbon electrode sputtered with palladium and gold for the amperometric detection of hydrogen peroxide. *Anal.Chim.Acta,* **178**, 247-254.

[11] Tarasevich, M.R., Radyushkina, K.A., Bogdanovskaya, V.A. (1991). *Electrochemistry of Porphyrins*, p. 197. Nauka Moscow.

[12] Bogdanovskaya, V.A., Horosova, E.G., Vorobiov, V.G., Tarasevich, M.R. et al. (1990). Enzymatic and electrochemical reactions with peroxidase. *Electrochem.* (Russia), **26**, 573-579.

[13] Wollenberger, U., Bogdanovskaya, V.A., Bobrin, S.V., Scheller, F. et al. (1990). Enzyme electrodes using bioelectrocatalytic reduction of hydrogen peroxide. *Anal.Lett.*, **23**(**10**), 1795-1808.

[14] Cardosi, M.F., Turner, A.P.F. (1991). Mediated electrochemistry: A practical approach to biosensing. *In: Advances in Biosensors 1* (A.P.F. Turner, Ed.), 125-169. JAI Press, London.

MECHANISM OF ELECTRON TRANSPORT BETWEEN REDOX PROTEINS, ENZYMES, AND ELECTRODES:
BIOSENSORS BASED ON MEDIATORLESS ELECTRON TRANSPORT

A. I. Yaropolov and B. A. Kuznetsov

OUTLINE

Advances in Biosensors
Volume 3, pages 31-52.
Copyright © 1995 by JAI Press Inc.
All rights of reproduction in any form reserved.
ISBN:1-55938-535-9

ABSTRACT

The mechanism of electron transfer from protein electron carrier to metal electrodes has been studied, and the role of protein adsorption is considered. The mechanisms of electron transfer at modified electrodes were discussed involving an unusual mechanism studied in which the rotational mobility of proteins in the adsorbed layer promotes a higher rate of electron transfer. Mediatorless bioelectrocatalysis was shown to proceed only in the case, when the enzyme possesses special properties. Besides the orientation of the enzyme on the electrode surface and the depth of location of the enzyme active site, the other important characteristics are the catalytic mechanism and the sequence of the substrate binding. In this connection the chemical reactions of hydrogenase, peroxidase and laccase have been studied. No correlation was found between the rate of electron exchange between enzymes and electrodes and their electrocatalytic activity. Approaches to the developement of biosensors based on mediatorless enzymatic electrocatalysis are considered.

1. INTRODUCTION

It is known that through evolution protein electron carriers have acquired a specific structure due to which they are able to exchange electrons with "external objects." Specificity of electron transfer and mutual recognition of the objects are provided by binding sites. When two natural partners interact with each other, the binding sites secure the proper orientation of the partners for the tunneling of electrons. The functioning of an enzyme supposes the complementary interaction of the enzyme with substrates. The electron transfer occurs within the enzyme-substrate complex and is determined by the mechanism of the enzyme action. The electron transfer to "external" objects is mediated by low molecular weight electron carriers or cofactors. Therefore the mechanism of electron exchange of an enzyme at an electrode and electron transfer in an electrocatalytic reaction are expected to differ from that of protein electron carriers. If for the latter it is sufficient to accomplish an appropriate orientation of the protein on an electrode surface with the help of functional groups of various compounds, this is not the case with redox enzymes and much of the behavior of redox enzymes is determined by the mechanism of the catalytic process. In some cases the lack of the second substrate results in a change of the limiting stage of electron transfer through a protein molecule.

In this review we present results obtained in our laboratories for the period from 1978 to 1992. The review comprises the problems of electron exchange between protein electron carriers and electrodes, the coupling of enzymatic and electro-chemical reactions, and the development of mediatorless biosensors.

2. SPECIFIC FEATURES OF ELECTROCHEMICAL REACTIONS OF PROTEINS

When studying proteins by electrochemical methods, special attention should be given to their conformational changes at the electrode/solution interface. Although the phenomenon of the protein unfolding on metals, polymers, and water has been known for more than 50 years [1], the rates of unfolding on the metal surface have not been determined to date. We succeeded in determining experimentally the lower limit of this rate at a mercury surface ($K > 10$ s^{-1}) [2]. The motive force of the protein unfolding is the surface tension at the interface. The surface tension on gold and platinum surfaces is significantly higher than that on a mercury surface ($\sigma = 426$ mN/m). Therefore, we would expect a higher rate of protein unfolding at the interface of these metals.

The surface tension of oxides of metals, polymers, and carbonaceous materials is much less. On these surfaces the adsorbed film can contain in the first layer both unfolded and native protein molecules. The denaturation completely ceases on hydrophilic surfaces (like dextran) at $\sigma < 1$ mN/m. An unfolded protein molecule is a planar tightly packed conformer looking like a disk, in which the secondary structure is partially retained. A monolayer of an irreversible adsorbed unfolded protein is a close package of such disks. It is obvious that there are pores between the disks. We proved the existence of pores in the monolayer by several methods and estimated their size [3-4]. Molecules of mediators and native proteins are able to discharge in the pores of the monolayer.

Conformers are able to expand with decreasing surface pressure, which results in further destruction of the secondary structure. The compression and tension of conformers are reversible processes [5].

On flattening, intraglobular groups of proteins may come into contact with an electrode surface and be registered by electrochemical methods. As a result, the redox potential of these groups changes significantly [4]. One should discriminate between electrochemical reactions of denatured molecules located in the first layer and the reactions of native molecules occurring in the pores of the monolayer. Redox reactions of the cysteine sulfur and prosthetic groups proceed in a layer of the denatured protein absorbed on a mercury surface. The reactions of native molecules in the pores of the monolayer proceed very slowly. The limiting stage is desorption from the pores ($K = 0.01$ s^{-1}) [6]. This mechanism is observed in electrochemical reactions of prostetic and disulfide groups of native molecules.

When the rate of electron exchange is high, as in the case of cytochrome c ($K \approx 10^3$ M^{-1} s^{-1}), native molecules adsorbed in the pores and in the second layer form an electrode modified with the reduced protein due to fast electron exchange (see scheme of an adsorbed layer in Figure 1).

Then the electron transfer of freely diffused molecules at the modified electrode turns out to be a fast process, and the diffusion of protein molecules becomes the limiting step [7]. However, not all the molecules diffused to the electrode are

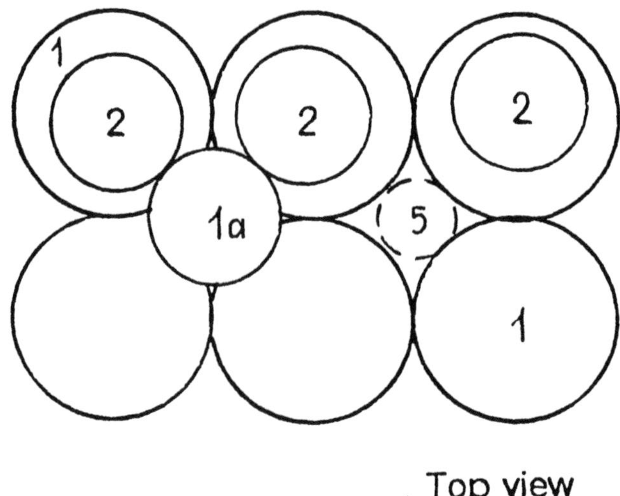

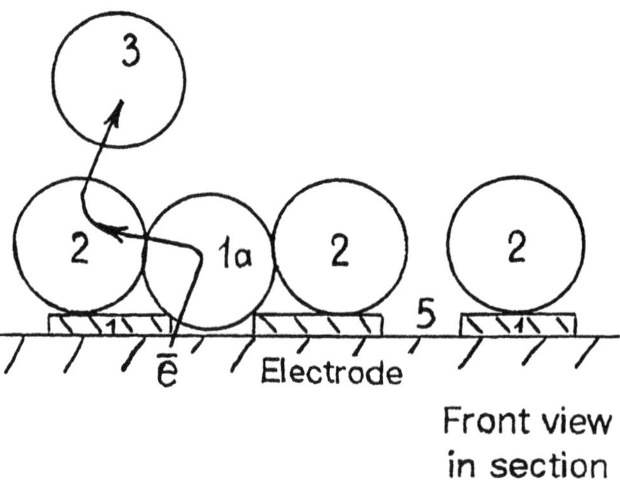

Figure 1. Scheme of the adsorbed layers and transfer of electron through: (1) the first layer of irreversibly adsorbed and flattened protein molecules; (1a) the native molecules adsorbed in the pores of the first layer; (2) the second layer consisting of the native molecules; (3) the freely diffusing molecules.

reduced at the potential of the limiting current, since the second stage is independent of the potential and determined by the rate of intermolecular electron exchange. This mechanism also explains other peculiarities of the cytochrome c electroreduction at a mercury electrode. The dependence of the overvoltage of the potential of the halfwave ($\delta E_{1/2}$) on the concentration of protein (C) derived on the basis of this mechanism is well satisfied [7]:

$$\delta E_{1/2} = -\frac{RT\Gamma_1}{\alpha n F}(S - S_0)\ln(1 + \beta c)$$

Here S_0 is the area of a pore; S is the area of a native molecule; B is the Langmuir constant of the adsorption isotherm of cytochrome c in the pores of the monolayer ; Γ_1 is the number of pores per cm^2.

Fast electrochemical reactions of multiheme proteins such as cytochrome c at a mercury electrode may be also accompanied by very quick denaturation. In a flattened conformer the heme groups are located very close to each other, thus providing electron transfer through the matrix of the protein adsorbed [8].

To perform electrochemical reactions of proteins on metal electrodes, the surface of the latter must be modified. This is necessary, to (1) reduce the surface tension and prevent denaturation of protein and blocking of the electrode surface, and (2) to provide the optimal binding of proteins with an orientation favorable for the tunneling of electrons. Promoters for electrochemical reactions of proteins were discovered by Eddowes and Hill [9], who described in outline the principle of the promoter action. Promoters are bifunctional compounds forming a bridge linkage between the functional groups of a protein and a metal. The first efficient promoter 4,4'-bipyridyl, was later, found not to stop the adsorption and denaturation of proteins [10]. However, the amount of the promoter located in the pores of the monolayer provides fast electron transfer from cytochromes c and b. The discharge of cytochrome P450 only occurs at high overvoltage [11]. More appropriate promoters were proposed by Taniguchi et al. [12]. They are bifunctional sulfur-containing compounds, which as was shown by surface enhanced Raman scattering by Niki et al. [13] are irreversibly adsorbed on a silver surface and cannot be replaced with protein.

One of the problems that attracted our special attention was the mechanism of acceleration of electron exchange in proteins at modified electrodes. It was also interesting to establish the role of biologically important binding sites of proteins and to elucidate whether the reaction of the electron exchange between the protein and the electrode may simulate electron transfer in biological membranes. The fact of the matter is that there are a number of cases of accelerated electron exchange in proteins, which can be most easily explained by the optimal orientation of the protein at the electrode on interaction of biologically important binding sites with the corresponding functional groups of the modifying monolayer. Among them are the well-studied reactions of cytochrome c at a gold DPDS-modified electrode and of plastocyanin at a cysteamine-modified gold electrode [14,15]. Here hydrogen bonds and electrostatic interactions provide a suitable orientation of protein. We found that the reaction of HIPISP from purple bacterial—a protein with a hydrophobic binding site—only occurs at an electrode modified with a hydrophobic compound ß-mercaptoethanol [15]. In the case of cytochrome c 553, whose binding site contains a mosaic of charges, the best results were obtained at a gold electrode modified with a mixture of thioglycolic acid and cysteamine [15].

Although the list of examples can be continued, it is noteworthy that there are a number of facts that contradict the above stated simple scheme. For instance, the electron exchange of cytochrome c at an electrode surface containing carboxyl groups proceeds very slowly or even does not take place, although a strong interaction between numerous lysine residues of the binding site and the carboxyl groups can be expected [15]. There is no electron transfer from plastocyanin to a gold electrode modified with cysteamine at pH < 7. Finally, it is impossible to explain in the framework of the given scheme the existence of the universal modifier ß-mercaptoethanol, which is able to accelerate with equal efficiency the electron transfer from proteins containing binding sites of different structure: positively and negatively charged, neutral, and hydrophilic.

In this connection we studied in detail the electron exchange reaction at a gold electrode modified with various compounds: cysteamine (CA), thioglycolic acid (TGA) and their mixture, ß-mercaptoethanol (ß-ME), and dipyridyl disulfide (DPDS). The results are presented in Table 1. The scanning of the modifying functional groups on the electrode surface enables the estimation of the contribution of hydrogen bonds, and hydrophobic and electrostatic interaction.

Besides, it was also important to determine the effect of the energy of interaction on the rate of electron exchange. It is known that ionization, for instance, of the NH_2-groups of CA in the adsorbed monolayer is significantly inhibited due to the reciprocal influence of neighboring charged groups [16]. According to our calculations, the apparent pK shifts by 5-7 units. As a result, a significant change of the surface charge density occurs over a wide range of pH, enabling the energy of the interaction between the plastocyanin binding site and the electrode surface to be changed by varying pH.

The electron exchange between plastocyanin and modified electrodes was studied by the voltammetric method described earlier in [15].

Figure 2 presents voltammograms of plastocyanin at a gold electrode modified with CA at various pH. It is clear that the heights of the cathodic and anodic peaks change significantly over the narrow range of pH 7-8, while their potentials remain unchanged.

Figure 3 presents the plots of the height of the cathodic peak versus pH for an electrode modified with CA, ß-ME, TGA, and CA-TGA mixture (1:1).

The analysis of the plots allows us to conclude that a sharp decrease of the rate of electron exchange between plastocyanin and a CA-modified electrode cannot result from conformational changes of the protein. This is because no changes are

Table 1. Rate Constants and Peak Currents of Electrochemical Reactions of Proteins at Gold Electrodes with Different Modifier

Protein	Plastocyanin		Cytochome c-553		HIPISP		Cytochrome c	
	$K \cdot 10^3$ cm/c	j $\frac{mA}{M \cdot cm^2}$	$K \cdot 10^3$ cm/c	j $\frac{mA}{M \cdot cm^2}$	$K \cdot 10^3$ cm/c	j $\frac{mA}{M \cdot cm^2}$	$K \cdot 10^3$ cm/c	j $\frac{mA}{M \cdot cm^2}$
Modifier								
Mercapto-ethanol	5.5	45	4.0	41	2.0	53	2.0	33
Cysteamine	0.86	18*	0.37	6	0	0	0	0
Thioglycolic acid	0	0	0	0	0	0	0	0
Cysteamine and thioglycolic acid	0.3	39	0.50	29	0	0	1.0	29
Di(4-pyridyl) disulfide	0	0	0	0	0	0	1.1	32
Cystein	–	–	0.15	31	0	0	0.56	30

Notes: $j = I/cq$, I - cathode peak current of voltammogram, c - concentration of protein, q - area of the electrode surface.
* This measurement was carried out at pH 7.6

observed during the electrochemical reaction of plastocyanin at electrodes modified with all the other compounds. On the other hand, the adsorption of CA cannot reach the degree which enables the explanation of the sharp decrease of current at pH < 7. We performed special experiments to establish the degree of CA desorption on voltammetric measurements. We naturally paid special attention to peculiarities of protonization of amino groups in the monolayer. By our estimates, the degree of protonization of amino groups on the electrode surface is still low at pH 8 (pK_{ap} 5-6), but with decreasing pH from 8 to 7 it significantly increases, thus providing the enhancement of the electrostatic interactions between plastocyanin molecules and the electrode surface and a considerable increase of the adsorption equilibrium constant.

We suppose that at pH 7 a firm binding and orientation of plastocyanin molecules takes place. The binding seems to occur via the two blocks of amino acids (42-44 and 59-61), each carrying three negative charges. The tunneling of electrons through Tyr-83 being rather long and slow, was undetermined by voltammetry.

Another orientation of molecules seems to be required for a fast electron transfer, i.e., the area with His-87 must face the electrode surface [17]. We assume that such orientation can be reached. If the adsorbed protein molecules are somewhat mobile, when these distances become shorter at a certain instant, it is obvious that the mobility is reached at a rather low density of the surface charge.

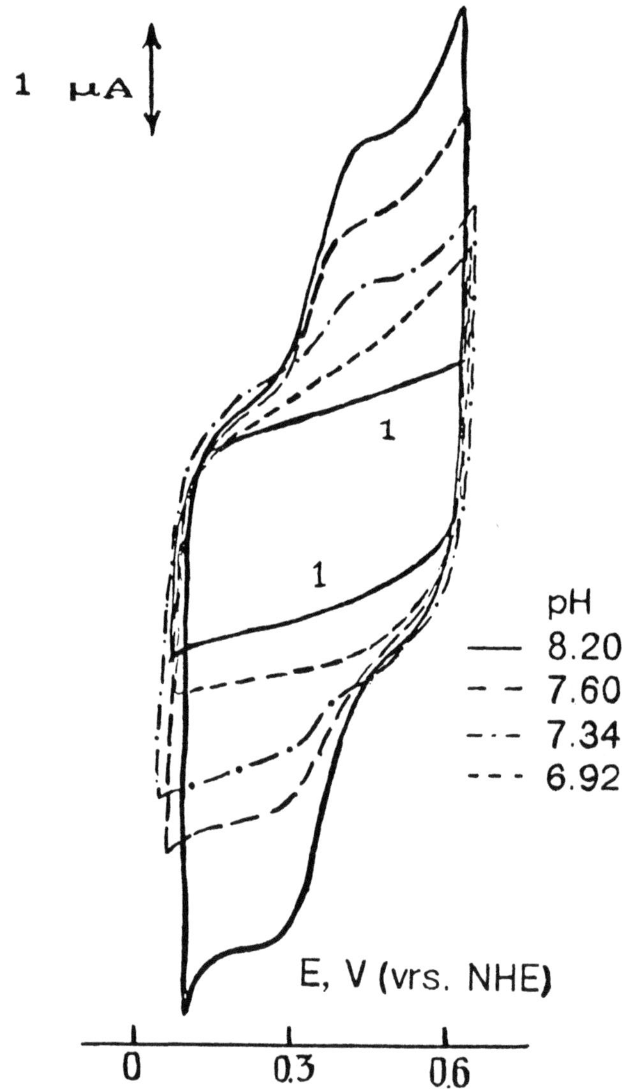

Figure 2. Voltammograms of plastocyanin (0.2 mM) at a gold electrode modified with cysteamine at pH shown in the figure and voltammogram of background electrolyte (1).

The rotational mobility of the protein molecules interacting with the electrode surface shows itself rather clearly in electrochemical reactions of various proteins at electrodes modified with ß-ME and a mixture of CA and TGA (1:1). The

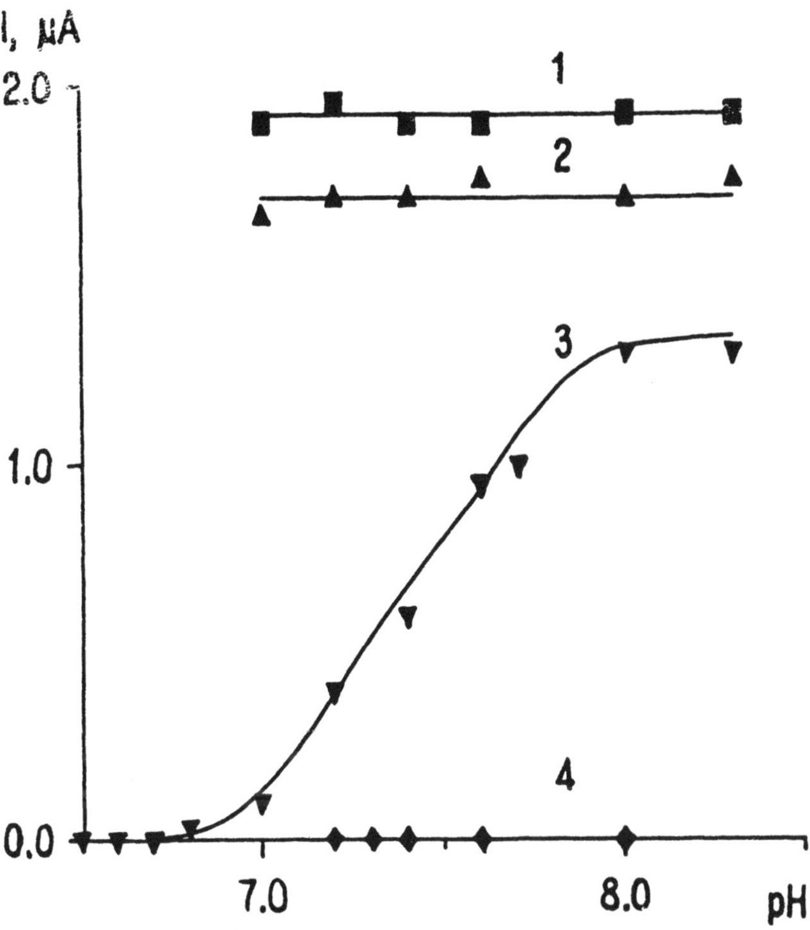

Figure 3. The anodic peak current at plastocyanine as function of pH at gold electrodes modified with (**1**) ß-mercaptoethanol; (**2**) composite of thioglycolic acid and cysteamine (1:1); (**3**) cysteamine; and (**4**) thioglycolic acid.

energy of the hydrogen bond between the hydroxyl group of ß-ME and the ionized carboxyl group and protonated amino group is the same and equal to 40 kJ/mol. The hydrophobicity of this compound is about 4 kJ/mol. This provides equal possibilities for the binding of all parts of the protein globule. Apparently the energy equivalency of the whole protein surface on interaction with the given modifier makes possible the rotational mobility of protein molecules.

As follows from Table 1, the electron exchange between plastocyanin and other proteins and the electrode surface proceeds with the highest rates at an electrode modified with ß-ME.

These notions of the role of mobility of protein molecules contacting with the electrode surface were also proved by other investigators [14]. For instance, the covalent binding of the cytochrome c favorably orientated to the surface of a glassy carbon electrode is an irreversible process [18]. The irreversibility of the redox reaction of plastocyanin at modified electrodes increases, if the protein is covalently bound to cytochrome c. When amino acids 42–44 and 59–61 are blocked with, for instance, cytochrome c, they cannot take part in "rigid" binding, and the binding can be performed in other directions [19]. Thus in the electrochemical process at a modified electrode, a firm orientated binding may be a hindrance to electron transfer, since electron exchange between proteins and electrode surfaces has specific features: the direction of binding and the direction of the optimal tunneling of electrons must coincide or be very close. In biological processes, this is unnecessary. In contrast, the orientated binding of the protein electron carrier with the membrane multienzyme complex occurs in special sites, which can be located at any distance from the inlet of the electron tunnel.

3. MEDIATORLESS ELECTROCATALYSIS WITH REDOX ENZYMES

The coupling of the enzymatic and electrochemical reactions can be realized using low molecular weight electron carriers and by the direct electron exchange between the enzyme and the protein.

The mediator coupling can be most easily performed. It is described in detail in reviews [20, 21]. Of great interest is the direct coupling of the enzymatic and electrochemical reactions without any low molecular weight electron carriers. In this case, the electrode is one of the substrates of the enzymatic reaction and serves as either a donor or an acceptor of electrons. Several such systems are described in the literature [22-27].

Let us consider the possibility of application of electrochemical kinetic methods to elucidate the mechanism of the reactions, using the hydrogen oxidation reaction on a carbon black electrode with immobilized hydrogenase as an example [22]. Hydrogenase catalyzes the oxidation of hydrogen with viologenes and cytochromes and the reverse reaction of the hydrogen formation from reduced carriers. It is known that electrodes made of carbonaceous materials are inactive in the reaction of hydrogen electrooxidation. A stationary potential of 0.00 is established at carbonaceous electrodes with immobilized hydrogenase in neutral solutions saturated with hydrogen, and the reaction of electrooxidation (electroreduction) of hydrogen accelerates (Figure 4).

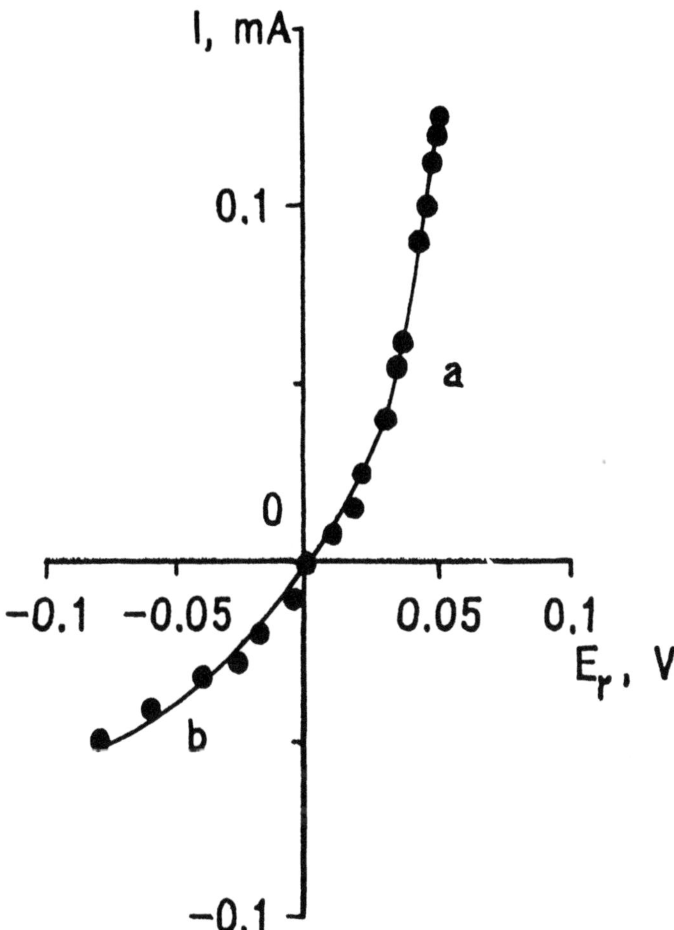

Figure 4. Polarization curve of hydrogen electrooxidation (**a**) and reduction (**b**) on an electrode with immobilized hydrogenase.

Some experiments were performed to elucidate the nature of the observed potential. In an argon atmosphere, the stationary potential shifted towards positive values, while in the presence of inactivated hydrogenase the background reaction of hydrogen electrooxidation in the observed region of potentials was virtually absent. The addition to the system of an inhibitor of hydrogenase, carbon oxide, also resulted in a complete disappearance of electrochemical activity of the enzyme electrode in the hydrogen atmosphere. Therefore, the potential observed is not associated with redox reactions of any groups of the protein and only established in the presence of hydrogen. A necessary condition is the presence of the active enzyme. The dependences of the stationary potential on the pH of the solution and concentration of hydrogen obey the Nernst equation and correspond to that of the equilibrium hydrogen potential.

The initial part of the polarization curve showing a linear dependence of overvoltage on current was used to determine the stoichiometric number, which indicates how many times the slow electrochemical step occurs during the catalytic process. The calculated mean value of the stoichiometric number is $1.0(\pm 0.2)$, which indicates that the slow step in the hydrogen ionization reaction occurs only once. This allows us to confine the number of possible mechanisms.

The polarization curves of hydrogen electrooxidation with semilogarithmic coordinates show two slopes. The mechanism of the reaction was analyzed in the potential range corresponding to the slope of the curve in the E, logI coordinate equal to $45(\pm 5)$mV. Taking into account the dependences of the reaction rate on pH and hydrogen partial pressure, a scheme can be proposed in which the transfer of the second electron from a hydrogen molecule to an electrode is the limiting step.

A more detailed analysis of the experimental results using the approaches of electrochemical and enzymatic kinetics and of the reverse reaction—hydrogen electroformation at an enzyme electrode—was made [23]. Hydrogenase was found to be reversibly inactivated at negative potentials due to the reduction of the functional group of the enzyme.

A characteristic feature of such a system differing from the systems with low molecular weight mediators is the possibility to regulate the rate of the enzymatic reaction by an electric field. Another example of mediatorless bioelectrocatalysis is the reaction of electrochemical reduction of hydrogen peroxide in the presence of immobilized peroxidase [25]. The enzyme is a typical representative of the group of redox enzymes containing hemin as a prosthetic group.

The electroreduction of hydrogen peroxide with peroxidase under anaerobic conditions proceeds at electrodes made of various materials (gold, pyrographite, carbon black) and depends on the amount of the enzyme adsorbed. The maximum stationary potential of a carbon black electrode is 1.24 V.

We compared properties of the immobilized peroxidase in the reaction of hydrogen peroxide reduction at the electrode and in the reaction of o-dianizidine oxidation. After immobilization on a carbon black electrode, peroxidase retains about 50% of its initial activity. The plots of the activity of the immobilized enzyme versus pH and $[H_2O_2]$ in an electrochemical system and in the reaction of o-dianizidine oxidation have almost the same shape (Figures 5 and 6).

The decrease of the rate of the electrochemical reaction at a hydrogen peroxide concentration of more than 1 mM and at pH < 4.5 is associated with the decrease of the catalytic activity of peroxidase. The formation of an inactive complex between the enzyme and hydrogen peroxide results in deterioration of electro- chemical characteristics of the electrode with time, and on long incubation to a

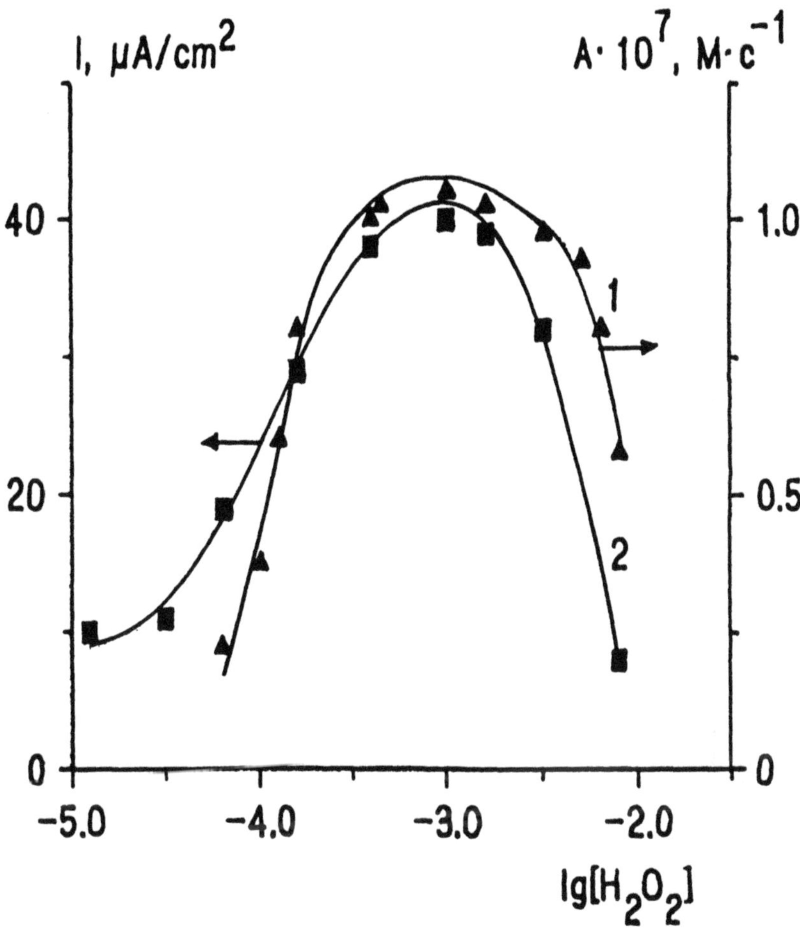

Figure 5. The dependence of the immobilized peroxidase activity on hydrogen peroxide concentration in the reactions: (1) oxidation of *o*-dianizidine; (2) electro-reduction of hydrogen peroxide at E_r = 1.1 V and pH = 5.0

complete loss of activity. The catalytic currents at a given potential in the presence of 1 mM H_2O_2 decreases to one-half in an hour. The stationary potential of the system reduces by 100 mV at 20 hours. The insufficient stability of the system with time associated with inactivation of peroxidase did not allow us to investigate in detail the mechanism of the electrochemical reaction. However, the results obtained indicate that the behavior of the heme-containing enzyme, peroxidase, in the electrochemical system is determined to a great extent by properties of the enzyme in the homogeneous state.

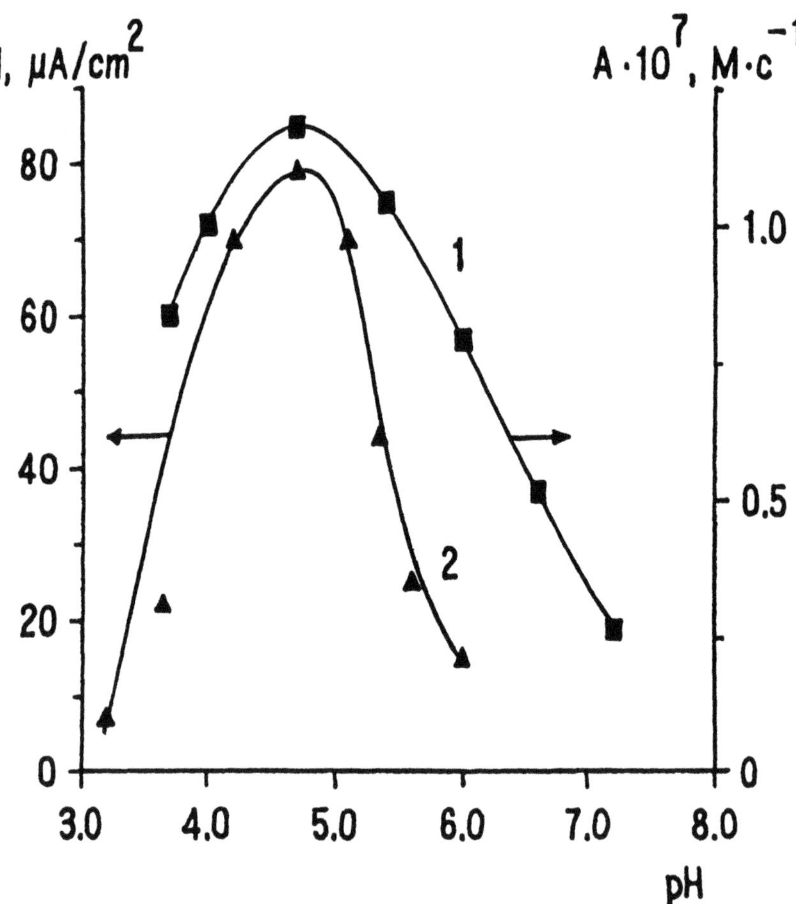

Figure 6. The dependence of the immobilized peroxidase activity on pH in reactions: (1) oxidation of o-dianizidine; (2) electroreduction of hydrogen peroxide at $E_r = 1.05$ V and hydrogen peroxide concentration 1 mM.

Mediatorless enzymatic catalysis of electrochemical reactions was shown for for the first time with oxygen electroreduction at various electrodes (gold, carbon black, pyrographite, glassy carbon) using immobilized laccase [24]. Laccase is a copper-containing enzyme catalyzing the oxidation of various electron donors with oxygen, which is reduced to water without the intermediate formation of hydrogen peroxide. The presence of laccase at the electrode surface accelerates the reaction of oxygen electroreduction over a wide range of potentials. On the basis of a detailed kinetic study of the enzymatic reaction in solution [28] and at electrodes, we proposed a mechanism of oxygen electroreduction in the presence of laccase.

The aforementioned effects of the mediatorless catalysis of electrochemical reaction with redox enzymes should be interpreted to elucidate the mechanism of the electron transfer between the enzyme and the electrode. From general considerations it is obvious that the less the distance from the electrode to the primary electron acceptor in the enzyme, the higher the probability of electron transfer.

The efficiency of electron transfer was studied [29, 30] for the reaction of oxygen electroreduction depending on the distance between the electrode and the electron acceptor in laccase. The distance was varied by adsorbing cholesterol from ethanol and heptane on a graphitized carbon black electrode and lecithin from benzene. As a result, lipid monolayers with different orientation to the electrode surface were formed. Of great importance is the fact that the adsorption of lipids on graphitized carbon black is studied rather well, which enabled us to use important information about the geometric characteristics of the system.

Laccase was adsorbed beyond a lipid monolayer. Depending on conditions, cholesterol can be adsorbed as a monolayer on the graphitized carbon black surface with two orientations: flat, when the "thickness" of the lipid layer is about 6Å, and vertical, when the lipid layer is 17 Å thick. Electrodes with a monolayer of adsorbed lecithin were also prepared.

Laccase was adsorbed at a graphitized carbon black electrode with a lipid orientated in this or another way to the electrode surface, and the rate of oxygen electroreduction was measured. The different specific activity of the enzyme on these carriers was taken into account by referring the currents of oxygen electroreduction to the maximum rate of the enzymatic reaction measured in independent experiments (Figure 7). The results obtained indicate that the efficiency of electrocatalysis with laccase depends on the distance between the enzyme and the electrode.

However, it is more correct to perform experiments with ordered lipid layers prepared by the Langmuir technique. The tunneling mechanism of electron transfer can help to explain the regularities observed and the effect of bioelectrocatalysis [29].

Summing up the results presented in this section it is noteworthy to emphasize the following: hydrogenase, peroxidase, and laccase can be used in electrochemical conversions of the corresponding substrates under conditions of direct (mediatorless) electron exchange between the electrode and the active site of the enzyme. The mediatorless catalysis proceeds both on gold and carbonaceous electrodes. This eliminates the assumption of the possible involvement in electrochemical reactions of the surface groups of carbonaceous materials having a quinone structure, which can act as mediators of electron transfer. For hydrogenase such groups cannot be a substrate at all. Finally, it is shown that equilibrium hydrogen and oxygen potentials are established at electrodes with immobilized hydrogenase and laccase instead of the potential of electron carriers (this is proved by the dependences of stationary potential on the concentration of substrate and reaction products.)

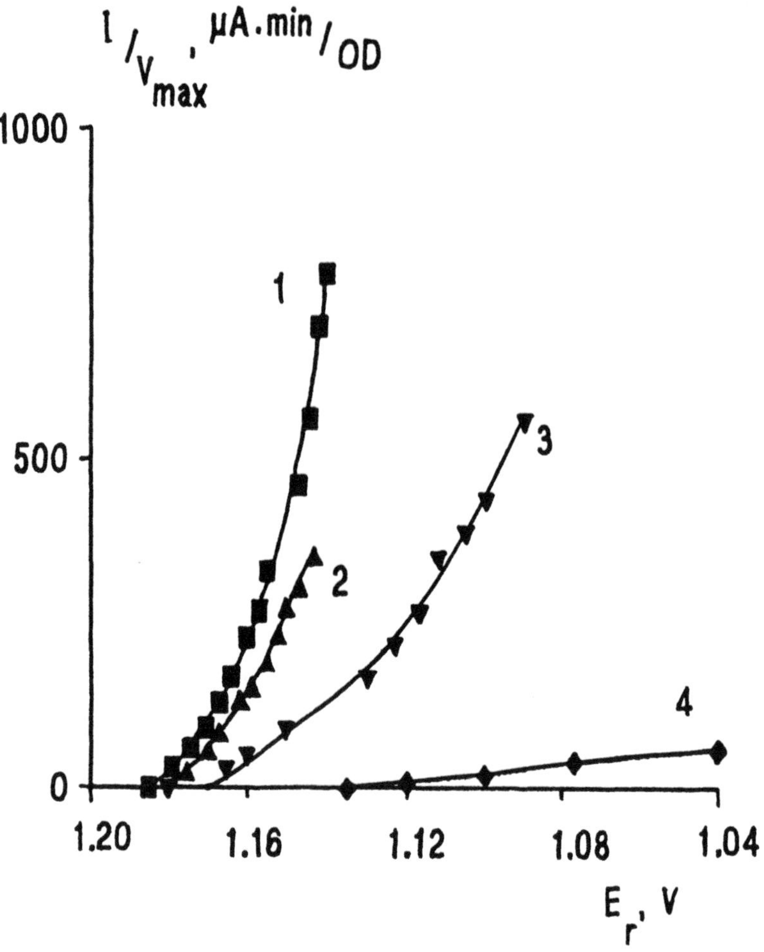

Figure 7. The dependence of the current referred to the electrode enzymatic activity on the potential for: (**1**) graphitized carbon black electrode; (**2**) graphitized carbon black electrode covered with cholesterol with a flat orientation; (**3**) with a vertical orientation; (**4**) graphitized carbon black electrode covered with leucitine.

4. ELECTROCHEMICAL REACTIONS OF ENZYMES AT ELECTRODES IN THE ABSENCE OF SUBSTRATES

To predict mediatorless bioelectrocatalysis, special attention should be given to studies on redox conversion of the enzyme prosthetic groups in the absence of substrate (noncatalytic process) at various electrodes, and to studies on the relations between these conversions and electrocatalytic properties of enzymes.

We studied the electrochemical behavior of metal-containing redox enzymes (laccase, ceruloplasmin, peroxidase, hydrogenase) at various electrodes under anaerobic conditions. Potentiodynamic studies of laccase and ceruloplasmin adsorbed at carbon black electrodes showed under anaerobic conditions the occurrence of a reversible electrochemical process. On electroreduction of proteins at potentials corresponding to the maximum currents, and simultaneous registration of spectral characteristics, we established the nature of the primary electron acceptor in the proteins. It was found to be copper type I. After reoxidation the enzymes retained completely their activity in oxidation of organic substrates [31].

Potentiodynamic and spectroelectrochemical measurements made at pyrographite, platinum, gold, and amalgamated gold electrodes with adsorbed peroxidase indicate that there is no electrochemical reaction of hemin in the protein [32].

As protons are a substrate for hydrogenase, potentiodynamic studies of the hydrogenase absorbed at an electrode in an atmosphere of an inert gas gave no unambiguous information about the possible occurrence of redox processes in the iron-sulfur cluster of the enzyme. The redox processes observed are attributed to the reaction of electrooxidation / electroformation of hydrogen catalyzed by hydrogenase [23].

The comparison of the results obtained in studies on redox conversions of the prosthetic groups of the aforementioned enzymes with those obtained in studies on their electrochemical properties gave firm evidence for the absence of a direct correlation between the redox conversions and electrocatalytic properties of the enzymes. Therefore, it is impossible to predict the catalytic properties of redox enzymes by their ability to exchange electrons with the electrode in the absence of substrates.

The bioelectrocatalytic effect is determined by a number of factors, each playing a greater or lesser role when studying a particular enzyme. The main factors are as follows. The order of the binding (activation) of substrates in any enzymatic reaction can be determined from the homogeneous kinetics. Apparently the electrocatalytic effect in the presence of an enzyme can be expected only for the substrate, which is the first to bind to the enzyme. Of great importance is the mechanism of catalysis with redox enzymes, i.e. it is necessary to elucidate whether the reaction proceeds with the formation of a triple complex or by the ping-pong mechanism. From general considerations, the acceleration by an enzyme of the electrochemical reaction proceeding without mediators should be expected only in the second case. This was proved, for instance, by kinetic and electrocatalytic studies with laccase and ceruloplasmin. Despite similarity of the sum reactions, the mechanism of catalysis has principal distinctions: laccase catalyzes the reactions by the ping-pong mechanism, while ceruloplasmin operates via the formation of the triple complex, donor-enzyme-oxygen. As a result, the latter fails to catalyze the reaction of oxygen electroreduction.

Among the other factors influencing the efficiency of the electrocatalytic process, the most important are the material of the electrode, the depth at which the enzyme active site is located in the protein globule, orientation of the enzyme on the electrode surface, and the existence of a certain contact between the enzyme and the electrode.

5. BIOSENSORS BASED ON ENZYMATIC ELECTROCATALYSIS IN THE ABSENCE OF MEDIATORS

The above-mentioned effect of mediatorless electrocatalysis may be realized in the development of biosensors for assaying various compounds. Peroxidase was used in a two-enzyme sensor with an amperometric detection system for glucose assay [33]. This method can be also used to detect other metabolites undergoing oxidation in the presence of oxidases which produce hydrogen peroxide. In some cases it is more convenient to make measurements in the potentiometric regime, since the stability of the potential change in time is significantly higher than the stability of currents. Mediatorless sensors for the detection of lactate [26] and fructose [27] are also described.

The ability of laccase to catalyse the reaction of mediatorless oxygen electrore- duction, and the possibility of performing an immunoassay with its participation, made a basis for the development of a novel type of sensor [34]. The principle of the action of such a sensor is shown in Figure 8.

The antigen to be assayed is covalently immobilized at a carbon electrode modified with a polymer. The addition to the reaction cell of the laccase-antibody conjugate results in the binding of the conjugate to the antigen on the electrode surface. This results in a rapid increase in the electrode potential due to media- torless catalysis by laccase of the electroreduction of oxygen.

The preliminary addition of the antigen to be assayed in the reaction cell causes a decrease in the rate of increase of the electrode potential as a result of competition between free and immobilized antigens for binding to the conjugate. Measuring the rate of change of the electrode potential at various concentrations of the antigen, we were able to plot a calibration curve. This method was used to detect human insulin and immunoglobulin. The sensitivity of the assay was 10^{-7} M. The assay duration was 20 min. It is noteworthy that the presence of the polymer enabled a partial elimination of nonspecific adsorption of proteins.

Despite the advantages of this immunoassay (the second substrate of the enzymatic reaction is not required), its sensitivity to the compound to be assayed is rather low. We have developed another method which is also based on the effect of catalysis by laccase in the absence of mediators. Registration was performed amperometrically after accumulation of charge in a double electrical layer for 10 min. This allowed us to increase sensitivity of the method up to 10^{-9} - 10^{-10} M.

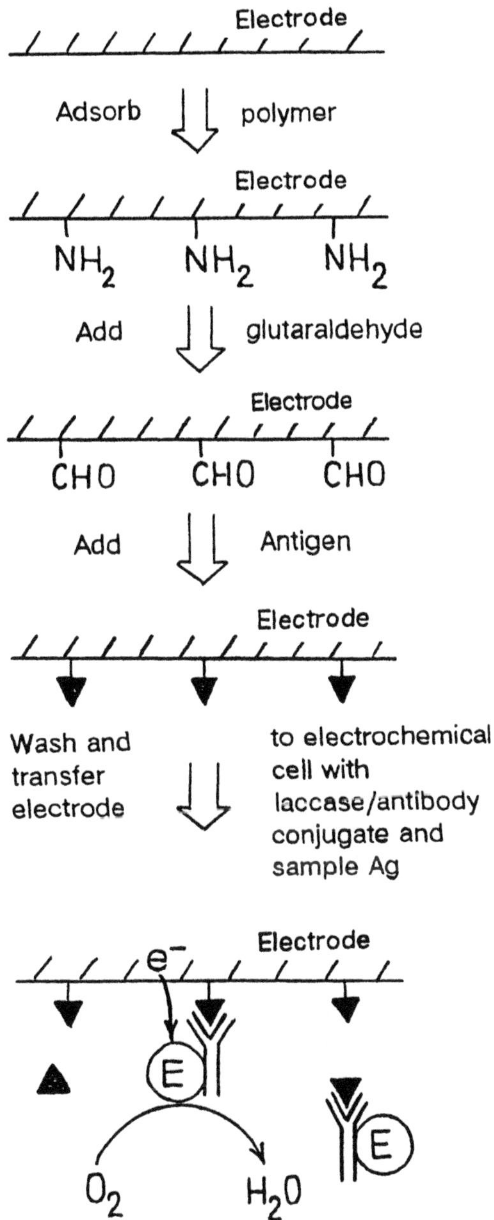

Figure 8. Diagrammatic representation of immunoassay using immunoelectrode based on the direct electrocatalytic reduction of oxygen in the presence of laccase. The electrode is coated with immobilized antigen, Ag. The antibody-laccase conjugate associates with immobilized antigen and catalyzes the reduction of oxygen.

Advantages of potentiometric sensors prepared by silicon technology can be combined with the advantages of laccase as an enzyme label in immunoassays. LAP-sensors can be used in the development of immunosensors for simultaneous assaying of many samples. The photocapacitance of such sensors depended on the pH of the solution in the oversurface region, or the redox potential of the solution in the case when a conductor is applied to a dielectric surface. A thin layer of gold or carbon was sprayed over a silicon plate with silicon nitride; the gold or carbon layer served as a working electrode. A nitrocellulose filter with immobilized laccase was pressed to the electrode. The reaction cell contained a solution of reduced ferrocene. On interaction with oxygen in the presence of laccase, ferrocene was oxidized and a change of the potential was registered at the electrode. Similar experiments were performed without ferrocene. The electrochemical reaction proceeded in this case by the mediatorless mechanism. Using this system, immunoassays can be performed both by competitive and sandwich methods.

REFERENCES

[1]   Adam, N.K. (1944). *The Physics and Chemistry of Surfaces,* 3rd edn. Oxford University Press, London, p. 436.

[2]   Shumakovich, G.P., Kuznetsov, B.A. (1984). Does a protein conformation change at the mercury electrode under conditions of polarographic analysis? *Electrokhimiya,* **20**, 448-454 (in Russian).

[3]   Shumakovich, G.P., Kuznetsov, B.A. (1979). On the porous structure of the protein monolayer adsorbed in the interface. *Biofizika,* **24**, 777-778 (in Russian).

[4]   Kuznetsov, B.A., Mestechkina, N.M., and Shumakovich, G.P. (1977). Electrochemical behavior of proteins containing coenzyme groups and metals. *Bioelectrochem. Bioenerg.,* **4**, 1-17.

[5]   Kuznetsov, B.A., Shumakovich, G.P. (1984). On conformational changes of proteins on the mercury electrode. *Electrokhimiya,* **20**, 147-153 (in Russian).

[6]   Kuznetsov, B.A. Shumakovich, G.P., Mestechkina, N.M. (1977). The reduction mechanism of cytochrome c and methemoglobin at the mercury electrode. *Bioelectrochem. Bioenerg.,* **4**, 512-521.

[7]   Kuznetsov, B.A. (1981). The mechanism of electrochemical reaction of hemoproteins. *Bioelectrochem. Bioenerg.,* **8**, 681-690.

[8]   Sagava T., Nakajima, S., Akutsu, H., Niki, K. (1991). Heterogeneous electron-transfer rate measurements of cytochrome c_3 at mercury electrodes. *J. Electroanal. Chem.,* **297**, 271-282.

[9]   Eddowes, M.J., Hill, H.A.O. (1977). Novel methode for the investigation of the electrochemistry of metalloproteins: cytochrome c. *J. Chem. Soc., Chem. Commun.,* p. 771-772.

[10] Kuznetsov, B.A. Shumakovich, G.P. and Mutuskin, A.A. (1985) Adsorption and electrochemical reactions of proteins at a mercury electrode covered by a monolayer of 4, 4' - bipyridyl. Bioelectrochem. Bioenerg., **14**: 347-356.

[11] Archakov, A.I., Kuznetsov, B.A., Izotov, M.I., Karulina, I.I. (1981). Electrochemical reduction of cytochrome P-450 in the presence of 4.4'-bipyridyl. *Dokl. AN SSSR.,* **258**, 216-219 (in Russian).

[12] Tanigushi, I., Toyosawa, K., Yamagushi, H., Yasukouchi, K. (1982). Voltammetric response of horse heart cytochrome c at a gold electrode in the presence of sulfur bridged bipyridines. J. *Electroanal. Chem.,* **140**, 187-193.

[13] Fan Ke-Yun, Satake, I., Ueda, K., Akutsu, H., et al. (1988). Redox Chemistry and Interfacial Behaviour Biological Molecules. (Drynhurst, G. and Niki, K. Eds.). Plenum Press, New York, London, pp. 125-138.

[14] Frew, E., Hill, H.A.O. (1988). Direct and indirect electron transfer between electrodes and redox proteins. Eur. *J. Biochem.*, **172**, 201-209.

[15] Kuznetsov, B.A. Shumakovich, G.P., Mutuskin, A.A. Mazhorova, L.E. (1990). Role of the binding sites of proteins in the electron transfer reaction both on the photosynthetic membrane and at the modified electrode. *Dokl. AN SSSR.*, **312**, 1244-1250 (in Russian).

[16] Solov'ev, A.A., Katz, E.Yu. (1990). Why is covalent immobilization of quinones possible at electrodes via aminosilanes? J. *Electroanal. Chem.*, **277**, 337-339.

[17] Brunshwig, B.S., Delaive, P.J., English, A., Goldberg, M., et al. (1985). Kinetics and mechanisms of electron transfer between blue copper proteins and electronically exited chromium and ruthenium polypiridine complexes. *J. Inorg. Chem.* **24**, 3743-3749.

[18] Kuznetsov, B.A., Byzova, N.A., and Shumakovich, G.P. (1993). The effect of orientation of cytochrome *c* molecules covalently attached to the electrode surface upon their electrochemical activity (in press).

[19] Bagby, S., Barker, P.D., Guo Liang-Hong, Hill, H.A.O. (1990). Direct electrochemistry of protein complexes involving cytochrom *c*, cytochrom *b*5 and plastocyanin. *Biochemistry*, **29**, 3213-3219.

[20] Wring, S.A., Hart, J.P. (1992). Chemicaly modified carbon-based electrodes and their application as electrochemical sensors for the analysis of biologically important compounds. *Analyst,* **117**, 1215-1229.

[21] Hill, H.A.O, Sanghera, G.S. (1990). Mediated amperometric enzyme electrodes In *The Practical Approach Series (Biosensors. A Practical Approach)*, (Cass, AT.G., Ed.), pp. 19-46. IRL Press, Oxford University Press.

[22] Yaropolov, A.I., Karyakin, A.A., Varfolomeev, S.D., Berezin, I.V. (1984). Mechanism of H2-electrooxidation with immobilized hydrogenase. *Bioelectrochem. Bioenerg.*, **12**, 267-277.

[23] Karyakin, A.A., Yaropolov, A.I. (1990). Electrochemical kinetics of hydrogenase from *Thiocapsa roseopersicina. Khimicheskay Fizika,* **9**, 1237-1243 (in Russian).

[24] Tarasevich, M.R., Yaropolov, A.I., Bogdanovskaya, V.A. Varfolomeev, S.D (1979). Electro-catalysis of a cathodic oxyden reduction by laccase. *Bioelectrochem. Bioenerg.*, **6**, 392-403.

[25] Yaropolov, A.I., Malovik, V., Varfolomeev, S.D., Berezin, I.V. (1979). Electroreduction of hydrogen peroxide on an electrode with immobilized peroxidase. *Dokl. AN SSSR,* **249**, 1399- 1401 (in Russian).

[26] Kulys, J.J., Sviemickas, G.-J.S. (1980). Reagentless lactate sensor based on cytochrome *b*2. *Anal. Chim. Acta,* **117**, 115-120.

[27] Khan, G.F., Kobatake, E., Shinohara, H., Ikariyama, Y., Aizawa, M. (1992). Molecular interface for an activity controlled enzyme electrode and its application for the determination of glucose. *Anal. Chem.* **64**, 1254-1258.

[28] Varfolomeev, S.D., Naki, A., Yaropolov, A.I., Berezin, I.V. (1985). Kineticks and mechanism of the catalytical reduction of molecular oxygen in the presence of laccase. *Biokhimiya,* **50**, 1411- 1419 (in Russian).

[29] Varfolomeev, S.D., Berezin, I.V. (1982). Bioelectrocatalysis, the Acceleration of Electrode Reactions with Enzymes. In *Advances in Physical Chemistry (Current Developments in Electrochemistry and Corrosion)*, pp. 60-95. (Kolotyrkin, Ya.M., Ed.), MIR Publishers, Moscow.

[30] Yaropolov, A.I., Sukhomlin, T.K., Karyakin A.A., Varfolomeev, S.D., Berezin, I.V. (1981). On a possibility of tunnel transfer of the electrone in enzymatic catalysis of electrochemical processes. *Dokl. AN SSSR,* **260**, 1192-1195 (in Russian).

[31] Yaropolov, A.I., Ghindilis, A.L. (1985). The relation between electrochemical properties of the prosthetic group and activity of laccase. *Biokhimiya* **21**, 982-983 (in Russian).

[32] Yaropolov, A.I., Tarasevich, M.R., Varfolomeev, S.D. (1978). Electrochemical properties of peroxide. *Bioelectrochem. Bioenerg.*, **5**, 18-24.

[33] Gorton, L., Jonsson-Petterson, G., Csoregi, E., Johansson, K., et al. (1992). Amperometric biosensor based on an apparent direct electron transfer between electrodes and immobilized peroxidase. *Analyst*, **117**, 1235-1241.

[34] Ghindilis, A.L., Skorobogat'ko, O.V., Yaropolov, A.I. (1991). Immunopotentiometric electrode based on bioelectrocatalysis in the absence of mediators. *Biomed. Sci.*, **2**, 520-522.

[35] Ghindilis, A.L., Yaropolov, A.I., Berezin, I.V. (1987). The role of the enzyme mechanism in manifestation of its electrocatalytic properties. *Dokl. AN SSSR*, **293**, 383-386 (in Russian).

DEVELOPMENT OF FET- AND LAPS-BASED BIOSENSORS

A. N. Reshetilov and S. M. Khomutov

OUTLINE

Advances in Biosensors
Volume 3, pages 53-76.

ABSTRACT

Three models of biosensors, based on pH-sensitive field-effect transistors (FET) and light-addressable potentiometric sensors (LAPS), are described in this review. The first model is a glucose biosensor. It uses cells of *Gluconobacter oxydans* immobilized in poly-*N*-vinylcaprolactam gel. The sensitivity of the sensor in sample solutions is within a concentration range of 3×10^{-5}-10^{-2} M. A biosensor for detection of human IgG makes use of the sandwich method of enzyme immunoassay. The calibration curve is linear within 10^{-10}-10^{-8} M IgG. The third model is a combination of FET and polymer membrane doped with photochromic dye spirobenzopyran. This approach is useful in constructing probe-type and even miniature sensors. Membrane was modified with nonactin to obtain a transducer sensitive to NH_4^+ ions. The responses are larger at low concentrations of NH_4^+ ions (10^{-6}-10^{-5} M) and smaller at high concentrations (10^{-3}-10^{-2} M). We also describe a method to measure pH distribution in agar nutrient media containing growing bacterial populations. Spatial pH profiles were obtained for populations expanding in peptone-containing media. The above models can be used in medicine, and biotechnology; the phototested sensor with spiropyran is the basis for development of miniature enzyme and cell biosensors.

1. INTRODUCTION

The last 10 years have witnessed intensive research and development in the field of biosensors—analytical devices in which a biological component is integrated with a physical transducer to detect or respond to specific interactions with chemicals. Biosensors use various types of physicochemical transducers: optical, acoustic, conductometric, calorimetric, and electrochemical. The type of transducer is determined by the biochemical reactions and conversions in the biosensor's receptor element. Many biosensor models, both commercially available and those described in research papers, are based upon semiconductor potentiometric devices including field-effect transistors (FET) and light-addressable potentiometric sensors (LAPS). Their advantages are small size and low cost in large-scale production; the membrane to immobilize a biological compound can be formed by methods compatible with semiconductor technology. The sensor and the system of processing the assay results can be arranged in one chip [1,2]. The most widespread at present are FETs sensitive to hydrogen ions; however, by forming a membrane in a certain way, they can be "tuned" to K^+, NH_4^+, Ca^{2+}, Na^+ and other ions. [1,3-5]. A new type of a semiconductor potentiometric device—the light-addressable potentiometric sensor (LAPS)—has been recently proposed by Molecular Devices Corporation (U.S.) [6,7]. LAPS is a further development of the idea of controlling the semiconductor state by external electric field. The authors aimed at creating a potentiometric transducer to simultaneously register several biochemical reactions without any noticeable complication of the design.

We believe that the obvious advantages of LAPS as a transducer for biosensor systems consist in the simplicity of its technological design combined with an absence of "drain" and "source" regions typical of the FET, as well as high potentiometric stability, an ability for simultaneous detection of many chemical events with a single device using a light beam instead of wire contacts. These facts taken together open up a possibility for wide use of LAPS for creating various types of biosensors. Below we consider in brief the possible uses of FET and LAPS in biosensor developments.

2. FIELD-EFFECT TRANSISTORS AND LIGHT-ADDRESSABLE POTENTIOMETRIC SENSORS AS A BASIS FOR BIOSENSORS

2.1 FET

pH-sensitive FET's are widely used for developing biosensors. In principle of operation and design, these transistors are virtually the same as field-effect transistors used in the electronic circuits of computers, TV sets, etc. The pH-sensitive field-effect transistor is based on direct conversion of the concentration of a chemical species into an electronic signal [1].

A biosensor based on a chemically sensitive FET (ChemFET) is schematically shown in Figure 1. The transistor consists of silicon substrate (p-type in this case) covered by an isolating layer of SiO_2. In a typical "electronic" transistor (the accepted designation is MOSFET, metal-oxide semiconductor) the surface of SiO_2 is covered by a thin layer of metal (gate electrode) to which potential U_G is applied to controll the transistor current. In the case of a ChemFET, which is the basis of a biosensor, a layer of either Si_3N_4, Ta_2O_5 or Al_2O_3 is formed on the surface of SiO_2 to improve the pH sensitivity. The ChemFET can be considered as a special type of "electronic" FET with the gate at a certain distance. Two n-silicon areas are known as the "source" and "drain." Current flows from the source to the drain through the channel of the transistor. A voltage U_G is applied to the ChemFET with respect to a reference electrode. To protect FETs from electrolyte they are encapsulated by chemically resistant varnishes or epoxy resins. Biological materials (enzymes, cells, antibodies/antigens) are immobilized on the surface of the dielectric by respective techniques. A voltage U_D is applied to the drain. The current I_D which flows from the source to the drain is measured as a function of U_G and U_D [1,2].

For MOSFETs, I_D depends on many parameters of the semiconductor, such as metal and silicon work functions, Fermi potential difference between the doped bulk silicon and intrinsic silicon, etc. In the case of the ChemFET, that is, when the metal gate electrode is replaced by the component "reference electrode-electrolyte," the above dependence of I_D is also supplemented by a dependence on pH. Then the drain current of the FET can be expressed in the following way (see [1,2] for details):

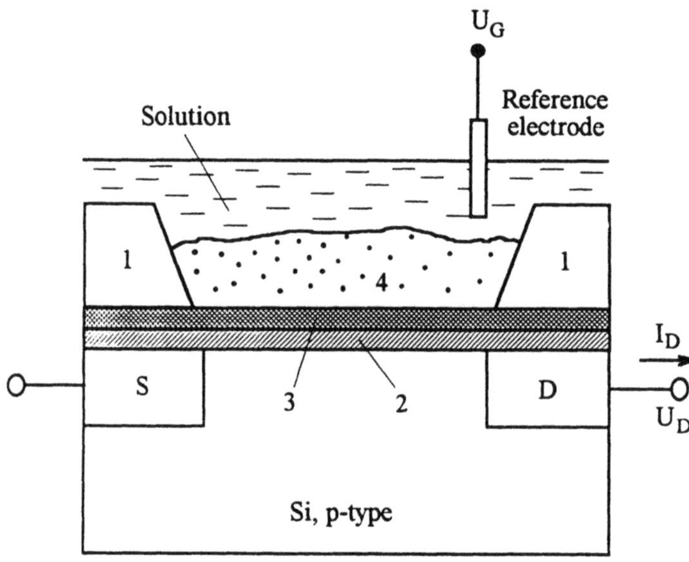

Figure 1. Schematic cross-section of a typical FET-based biosensor. (**1**) insulating encapsulant, (**2**) insulator, SiO_2, (**3**) chemically sensitive membrane, Ta_2O_5 (**4**) immobilized biomaterial. (S) source; (D) drain; (I_D) drain current; voltage U_D is applied to the drain, U_G to the reference electrode.

$$I_D = k \left(U_G - U_T - \alpha pH + \text{const}\right)^2 \qquad (1)$$

where the terms "const" and k contain parameters related to the semiconductor properties and depend on the concentration of protons in the medium, and α is a parameter that reflects the chemical sensitivity of gate insulator; U_T is the threshold voltage.

Equation (1) describing the pH dependence of the drain current on pH of solution shows how ChemFETs (a pH-sensitive FET, or pH-FET) can be used for the development of biosensors. To date, almost all biosensors based on pH-FETs use the enzymatic reactions where hydrogen ions are produced or consumed. Typical examples could be found in ref. [8].

Use of pH-sensitive FETs made it possible to develop diverse models of biosensors. A drawback of this type of biosensor is that its sensitivity depends on the buffering capacity of the solution. An increase in the buffering capacity brings down the value of local pH changes occurring in the bioreceptor area of the sensor, thus lowering the sensitivity. Buffer components, however, are a must sometimes to provide an optimal pH for biochemical reactions; in other cases the analyzed media themselves contain a buffering component (blood, urea, fermentation broth). An effective method to eliminate the effect of the buffering component is discussed in ref. [9]. A pH-FET-based urea sensor is combined with a noble-metal electrode which provides continuous coulometric titration of the products of an enzymatic reaction. The sensor thus becomes independent of the buffer capacity of the sample; because the enzyme operates at constant pH, the linear response range is expanded.

One of the new trends in the use of FETs is to develop transistors selective to various ions—for example, K^+, Na^+, Cl^-, and F^- [3-5]. Thus, the use of F^- selective transistor in a glucose or lactate biosensor makes it possible to avoid the effect of the buffering capacity of the solution [3].

Significant prospects in the field of biosensors are associated with the development of immunosensors possessing the highest sensitivity and selectivity. From the early 1980s, the FET began to be used for developing immunosensors based on the registration of changes in the electric charge of antibodies immobilized on the gate area during the interaction with antigens [10]. Since then, several research groups have tried to detect directly an immunological reaction by means of immunomodified FET. The detection of the charge redistribution caused by an immunological reaction has not been successful so far, because the charges are screened by small inorganic ions in the samples. A detailed analysis of this issue can be found in refs. [10-12]. At the same time, despite the early failures the search for the ways to carry out direct immunodetection continued. An original solution was proposed in a series of works by Bergveld et al. [11-14]. The authors described a new method for the detection of an immunological reaction in membrane. Antigens (or antibodies) are taken up in a porous membrane, which covers the gate area of an FET. Stepwise changes in electrolyte concentration evoke a transient transport of ions across the membrane-protein layer, thus generating a transient membrane potential to be measured by the FET. The transport is determined by the fixed charge density in the protein layer, which changes upon formation of antibody-antigen complexes. No membrane potential is induced at zero-charge density. Isoelectric points can therefore be deduced from the registration, and this forms the basis for sensing an immunological reaction. Sensitivity to a concentration of an antibody of 10^{-6} M can be easily achieved [13].

2.2 LAPS

Light-addressable potentiometric sensors are based on a silicon semiconductor substrate covered with a thin layer of dielectric material sensitive to hydrogen ions. Their principle of operation is similar to that of a transistor: at the electrolyte-insulator interface a surface potential is formed which affects generation of current in the semiconductor. However, in contrast with FET, LAPS are much simpler in design, they contain no "source" or "drain" regions, and the value measured is photocurrent generated in the semiconductor during its pulse illumination [6,7].

Attractive features of LAPS include potentiometric stability and ability to address different regions of the semiconductor by light rather than by wires. The ability to optically address different regions of a sensor surface allows multiple potentiometric measurements to be made with a single semiconductor device.

The significance of the new type of transducer is indicated by a number of publications which report diverse applications of the sensor. McKnabb et al. described a LAPS-based immunosensor for the assay of trace amounts of DNA [15]; LAPS was used in an immunofiltration procedure for the detection of pathogenic bacteria [16]; Rogers et al. described an enzyme electrode based on LAPS which contains immobilized acetylcholinesterase [17]. Another device called a microphysiometer, which makes it possible to register the intensity of living cell metabolism with high sensitivity, is described in ref. [7].

To sum up, LAPS is a promising new type of semiconductor transducer which will be used considerably more in the near future.

3.   RESULTS

In this section, we deal with the development of biosensors based on pH-sensitive FET and LAPS. Below some of the biosensor models developed in 1991-1992 are described.

3.1   Sensor Chips

The development of FETs is described in a number of works; however, each laboratory, as a rule, uses its own method and modification. We used a process similar to that described in [1,9]. The sensor chip contained one or two transistors. FETs were produced using NMOS technology on p-type silicon substrate. The silicon dioxide gate insulator (typically 80-120 nm thick) was formed by thermal oxidation of silicon surface. Ta_2O_5 (80-100 nm thick) was used as the second layer and the pH-sensitive gate insulator. The size of the gate was 40 μm x 500 μm. The size of the chips chosen were as large as 3 mm x 6 mm to facilitate encapsulation. The devices were then encapsulated with epoxy, covering all of the chip except the gate areas. The transistors had a channel resistance of about 1 kOhm and chemical sensitivity between 40 and 50 mV/pH. The leakage current in the circuit "reference electrode-silicon substrate" did not exceed 1 nA; the voltage drift and the noise were no more than 0.1 mV/h and 0.01 mV/h, respectively. The FET responses were measured in the constant gate voltage mode.

LAPS were fabricated by the same technology as FET. The sensors were round (6 mm in diameter); the diameter of the pH-sensitive zone, containing Ta_2O_5, was 2 mm.

The experiments with LAPS were done in two stages. First, an installation was set up, which consisted of commercial devices. A schematic representation of the

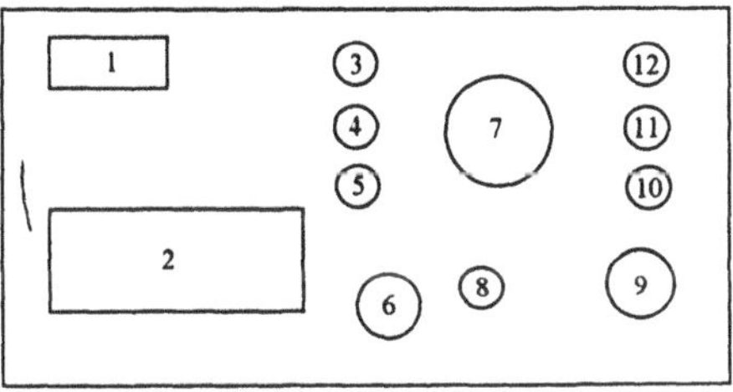

Figure 2. Schematic diagram of LAPS (**a**) and photocurrent as a function of bias potential (U_b) and solution pH (**b**) for used silicon sensors. The alternating photocurrent (I) is generated in the circuit under pulse illumination of semiconductor with light-emitting diode, LED. Photocurrent depends on U_b and also on surface proton concentration. Biochemical reactions occurring in the biomaterial immobilization zone change the local concentration of protons thus inducing a change in the output signal U_{out}.

sensor is shown in Figure 2a. Upon studying the properties of the sensor, we developed a measuring system and constructed a small-size LAPS device (Figure 3). The set-up is based on a combination of analogue and digital integrated circuits. The device makes it possible to register the dependence of the photocurrent on bias potential, to find the inflection point of a given dependence, as well as to record the local changes of pH in the zone of immobilized biomaterial. Typical dependences of photocurrent on bias potential obtained for solutions with

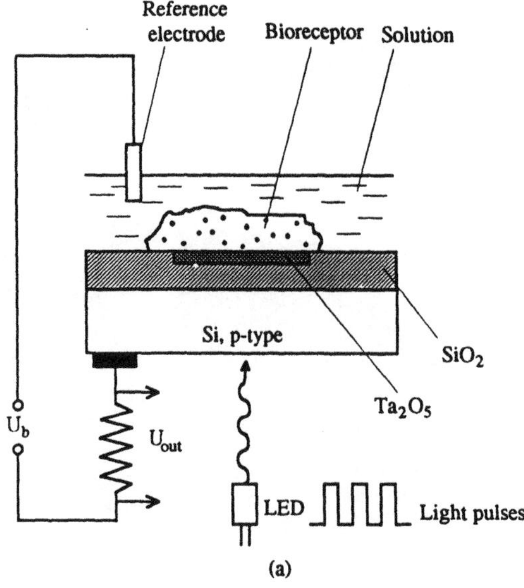

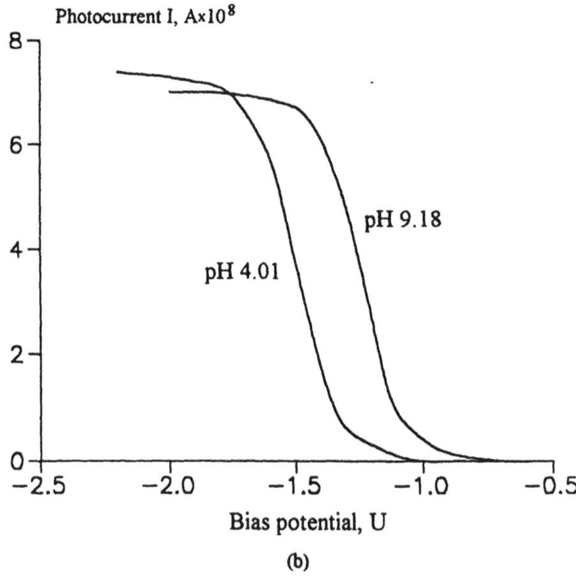

Figure 3. The appearance of the device developed: (**1**) name of the device; (**2**) digital indicator of output signals; (**3**), (**4**), (**5**) and (**7**) X-Y recorder knobs ("stop," "start," "reset," "sweep rate"); (**6**) operating current knob; (**8**) "manual-automatic" switch; (**9**) output jack (**10**), (**11**), (**12**) adjustment knobs ("frequency," "gain," "compensation").

different pH and reflecting the pH-sensitivity of the silicon sensors used are shown in Figure 2b.

3.2   Microbial Glucose Sensor with
Gluconobacter oxydans Immobilized Cells

When we planned experiments on biosensors, we tried to follow the most promising and significant trends in the biosensor field. With this in mind, we began experiments on sensors with immobilized microbial cells which are an example of a fast developing branch of biosensor technology. Immobilized whole cells are widely used now as receptor elements for biosensors. More than 50 organic compounds can be detected by microbial sensors [18].

Inexpensive glucose assays are of great interest in clinical analysis, the food industry, and fermentation control. So far a great number of different types of biosensors for glucose assay have been described based on enzymes or cells with amperometric or potentiometric detection [19-22]. The question arises whether it is worth developing a new glucose sensor. In our opinion, the answer is "yes" because not all the parameters in the described types of sensors (such as selectivity, sensitivity, stability, accuracy, cost, etc.) sufficiently satisfy the user. Thus, a glucose biosensor based on immobilized cells of *Gluconobacter oxydans* and a pH-sensitive field effect transistor has been developed.

Materials

Agar, yeast extract (Difco, U.S.), and tannin (Serva, Sweden) were used. Poly-*N*-vinylcaprolactam (PVNC) (MW 900,000) was obtained by *N*-vinylcaprolactam polymerization. Other chemicals were of reagent grade.

Microorganism and Cultivation

Gluconobacter oxydans str.6 was obtained from the All-Russian Collection, Institute of Biochemistry and Physiology of Microorganisms, and the Russian Academy of Sciences. Cells were maintained on sorbitol-yeast extract agar slants at 4 °C. The bacteria were grown aerobically by shaking (220 rpm) at 30 °C on sorbitol-yeast extract medium (pH 5.5-5.8). After 20-22 h of cultivation, the cells were centrifuged, washed twice with sterile tap water, and diluted with tap water to a 50 g/l cell suspension (dry weight/volume).

Immobilization Procedures

We tested various methods of cell immobilization and various types of gels: PVNC, zosterin, carrageenan, and agar. PVNC gel was the most effective; cells preserved their viability for a longer period. The entrapment of cells in PVNC gel membrane was performed according to a previously described method: cell suspension was mixed with stock solution of PVNC (5 -10%) and 10-15 µl of the mixture obtained was spread over the flat surface of FET and dried [23]. The PVNC-cell membrane obtained was stabilized by 0.5-1 % tannin heated at 38-40 °C over 3-5 min and washed twice with sterile tap water.

Measurements

FET-Based Model. Figure 4 shows the calibration curve and typical responses of the sensor to glucose addition into the cuvette (inset). A short time of about 1-2 min was necessary to stabilize the output signal of the sensor, so the measurement time was rather short, too. The sensitivity of the sensor allowed glucose to be determined within the range of $3 \times 10^{-5} - 10^{-2}$ M.

Substrate specificity is one of the most serious problems for sensors based on microbial cells. We chose this type of cells after screening several strains. All the compounds were tested at a concentration of 1.0 mM. Only two sugars—maltose and galactose—gave responses (correspondingly, 20 and 9% of glucose response)

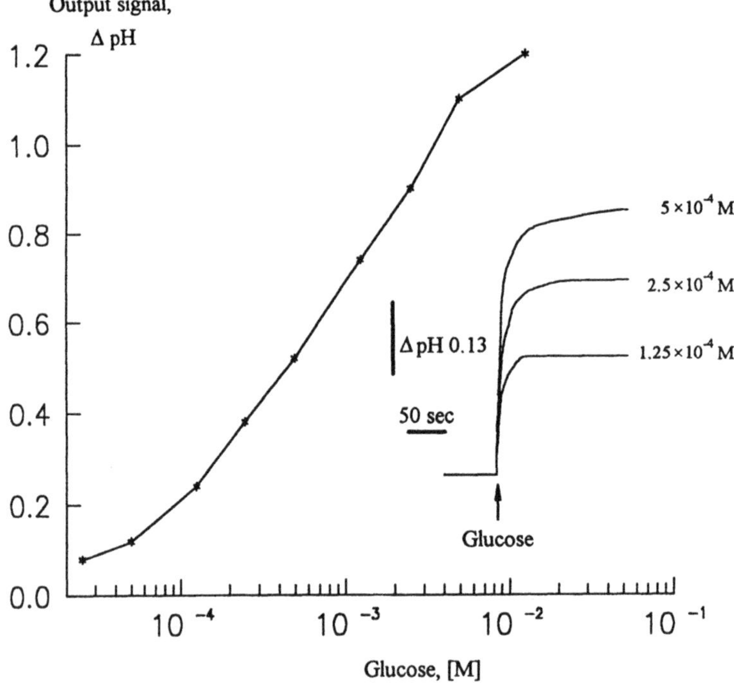

Figure 4. Calibration curve for FET containing immobilized cells of *Gluconobacter oxidans*. Measurement conditions: 15 mM NaCl, pH 7.2, no buffer components, temperature 22 °C. Output signal of the sensor is represented by changes of pH in the bioreceptor zone upon addition of glucose. *Inset*: typical sensor responses to the appearance of glucose in the medium.

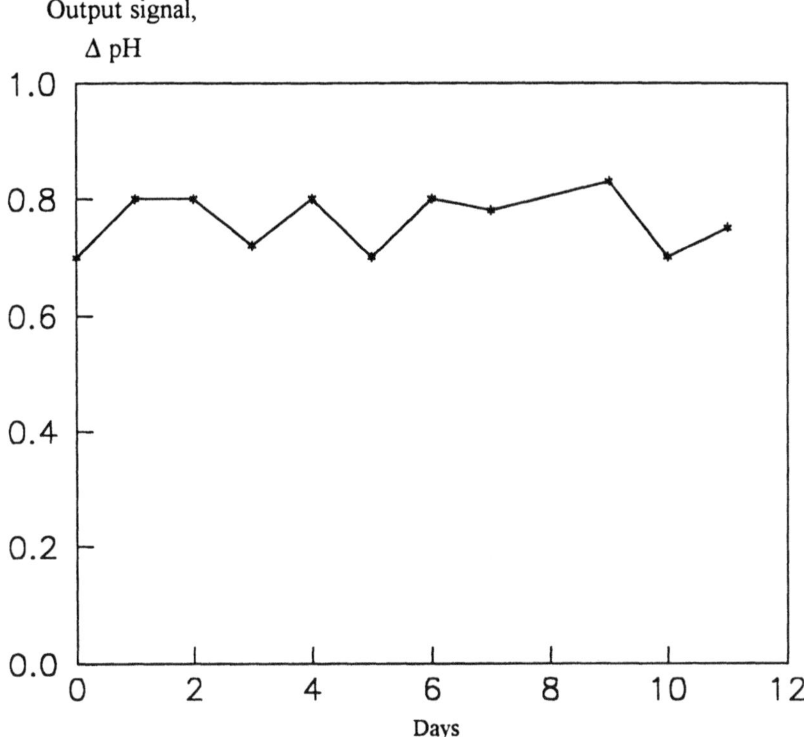

Figure 5. Stability of the FET-based model. 10-15 measurements were carried out daily. No substantial changes were observed during this period. Storage conditions: 10 mM phosphate buffer at 4 °C, pH 7.2.

that should be taken into consideration; the signals from the others (xylose, arabinose, sorbose, ramnose, sucrose, fructose, lactose, sorbitol, mannitol, inositol, xylitol, dulcitol, glycerol) were very small or none at all.

The biosensor was analyzed for two types of stability: long-term and operational. After storage at 4 °C for 2 months the biosensor responses to glucose were close to the initial results. As to the operational stability which was tested in continuous operation, Figure 5 shows that the responses were stable for no less than 10 days.

LAPS-Based Model. In search of new capabilities of biosensors for glucose assay, we made the first step in the development of a biosensor electrode based on LAPS and immobilized microorganisms. To immobilize microorganisms on the sensor we used the same procedure as for immobilization on the transistor. The measurements were carried out using the device developed by us (see Figure 3).

Typical current-voltage characteristics of the sensor containing no cells are given in Figure 2b. The change in the concentration of protons in the medium

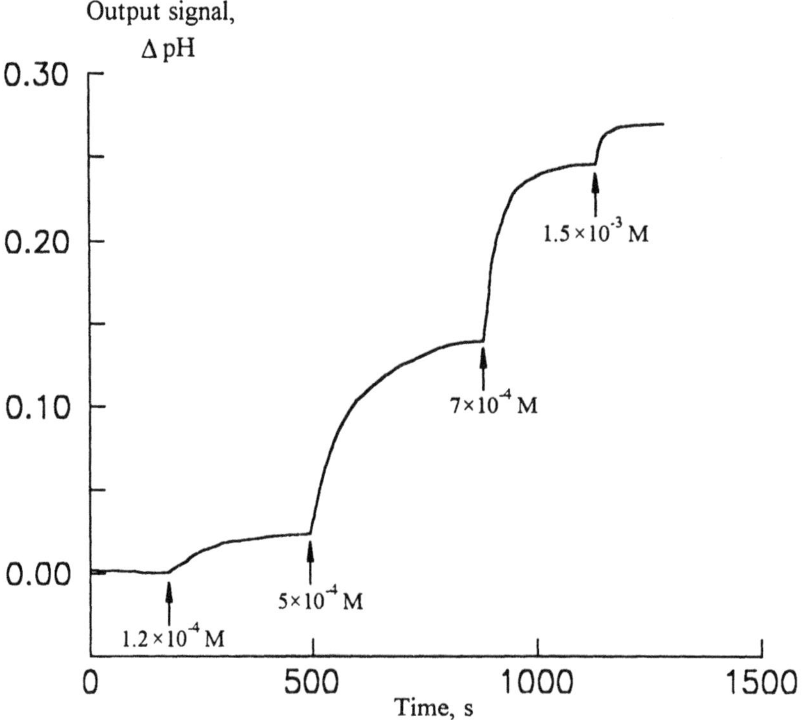

Figure 6.   LAPS with microbial cells as the basis of glucose sensor. Typical responses to the addition of glucose are shown. Output signals are expressed as pH changes in the immobilized cell layer. The arrows show the moments of glucose addition, the numbers indicate the final concentration in [M]. Measurement conditions are the same as in the legend to Figure 4. Frequency of light pulses, 2 kHz.

(change of solution pH) results in the generation of the electrochemical component of EMF at the dielectric-solution interface which shifts the current-voltage characteristics along the axis of the bias potential. This effect is the basis for using LAPS as a biosensor transducer. A typical response of biosensor to the successive increase in the concentration of glucose in the cuvette is shown in Figure 6.

The above models can be considered as a prototypes for commercial devices. The possible fields of application are fermentation technologies, food industry and medicine. To make the practical application possible, a number of problems are to be solved, one of which is associated with the dependence of the sensitivity of the sensor on the concentration of the buffer component in the media.

3.3 Immunobiosensor for IgG Detection

Construction

In recent years radioimmunoassay, fluoroimmunoassay, and enzyme immunoassay were included into the arsenal of the most widely used methods of immunochemical analysis [24]. One of the problems is to develop fast, safe and cost-effective methods. It is believed that the use of the biosensor approach can be useful in solving some of these problems—in particular in the development of compact and sensitive immunoassay devices for "on-site" or "doctor's office" analyses that are fast and independent of centralized biomedical services. Human immunoglobulin G (IgG) is an important indicator of abnormality in protein metabolism. Enzyme immunobiosensors have been developed for assaying IgG [25-28].

The IgG immunosensor described in ref. [28] was constructed using an immobilized human IgG on the polymeric membrane covering the gate region of pH-sensitive FET. The assay involves (1) competitive immunochemical reaction of urease-labeled antihuman IgG with human IgG in samples, and (2) membrane-bound IgG and electrochemical detection of membrane-bound urease activity. We have chosen the described model as a prototype and have developed our own approach by introducing some new elements. We used the sandwich method of enzyme immunoassay; besides, we applied a new technique of membrane fixation at the gate region of FET. In the method, we recovered the activity of the sensor after measurements by removing the membrane and using the new one. The membrane was used as a matrix to immobilize antibodies to human IgG.

Experimental

Human IgG, anti-human IgG, and horseradish-labeled antihuman IgG were purchased from "Immunotech" (a small business operation at Moscow State University). We used membranes from regenerated cellulose on nylon net KS-49 ("Polymersynthesis," Vladimir, Russia). The membrane prepared from regenerated cellulose had a porous structure and was about 80-100 micrometers thick. The membrane was impregnated with p-azidobenzaldehyde and illuminated with UVlight. The membrane fractions, 1.5 x 1.5 mm in size were fixed to the gate region of FET using a microholder. The developed system of fixing the membrane on the transistor did not prevent the supply of reagents to the membrane from the solution; besides, one can register with high efficiency the biochemical reactions occurring on the surface of the membrane and accompanied by a change in the concentration of protons. The FET with membrane was immersed for 10 min into the solution of polyclonal goat anti-human IgG (50 µg/ml). The protein was immobilized by the Schiff reaction of the aldehyde groups on the matrix with amino groups of immunoglobulins.

After the FET was washed to remove antibodies noncovalently bound to the membrane we blocked the surface by bovine serum albumin (1% solution, 30

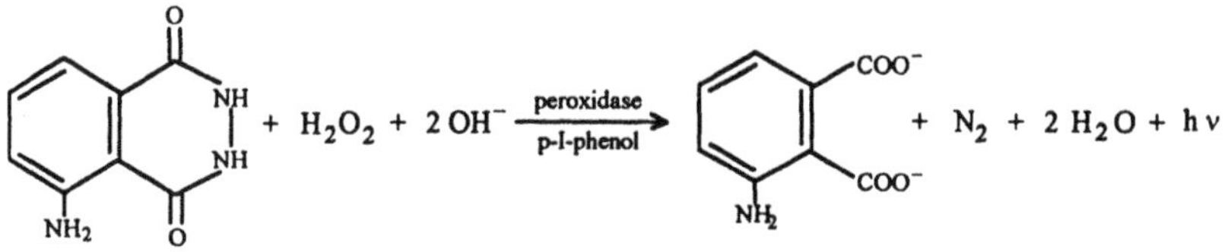

Scheme 1

min) to prevent nonspecific binding of proteins during subsequent stages. Next, the FET was immersed for 40 min into a test solution of IgG at concentrations from 10^{-11} to 10^{-6} M, and then into a solution of an antibody labeled with horseradish peroxidase (5×10^{-9} M , 50 min). Thus a sandwich was formed. At the final stage after the washout of unbound proteins we added the substrate mixture into the cuvette and recorded the pH shift. We used the same substrates as in the reaction of enhanced chemiluminescence, namely, p-I-phenol (5×10^{-5} M), luminol (4×10^{-5} M), and hydrogen peroxide (2×10^{-3} M) [29]. The pH changes were due to the appearance of 3-aminophthalate—the acid product of the reaction (Scheme 1).

The calibration curve of immunosensor is plotted in Figure 7. The measured parameter was the initial rate of pH changes, pH/t. The lower detection limit is about 10^{-11} M. For this point, the signal-to-background level ratio is equal to two units. The linear part of the calibration curve is from 10^{-10} to 10^{-8} M of IgG. The variation coefficient did not exceed 10% (n = 6). Thus the sensitivity of the sensor is rather high and gives us the opportunity to determine IgG at concentrations considerably lower than in human plasma, where the average concentration of this protein is about 10^{-5} M.

Figure 8 shows the pH dependence of sensor responses. The maximum values of responses are obtained in the pH range of 8.3-9.0, which is in good agreement with the data on enhanced luminescence enzymatic reaction.

We used this model for the preliminary estimation of the IgG concentration in human plasma of patients with normal immunological status. The obtained values of the IgG concentration varied from 10^{-5} to 5×10^{-5} M. We did not verify the data obtained by alternative methods so these results should be considered as preliminary.

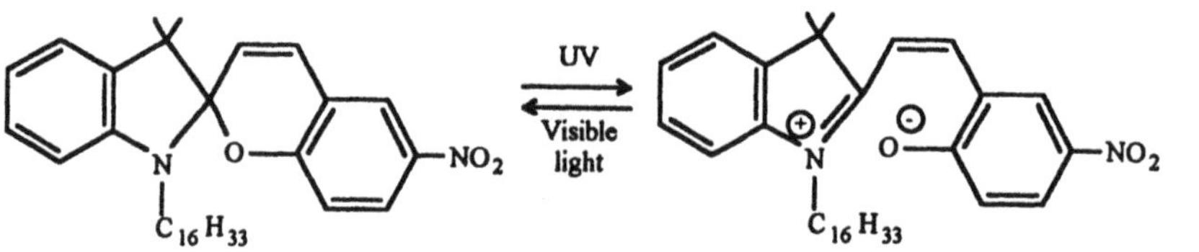

Scheme 2

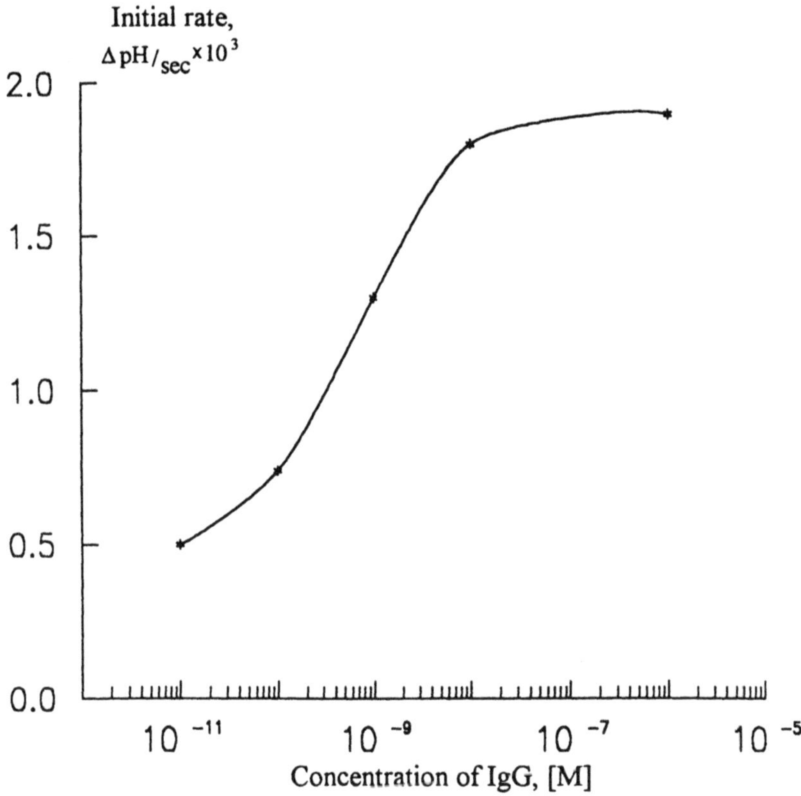

Figure 7. Calibration curve for human IgG detection. The measurement conditions: pH 8.7; 1 mM Tris-HCl buffer; 0.15 M NaCl; temperature 20 °C.

3.4 Phototested FET-Based Chemical Sensor

The next model is a photo-switchable or phototested ion sensor. Recently, much interest has been expressed in artificial membranes with controlled characteristics—for example, ion permeability, membrane potential, conductivity, etc. Photocontrol of membrane functions could be achieved by doping a membrane with a photosensitive compound. Many papers report on the use of spiropyranes subjected to photoinduced reversible transitions accompanied by color change and charge separation [32,33] (see Scheme 2).

This phenomenon is used in photomemory elements, phototransducers, and analytical sensors [30-34]. An essential feature of a new class of ion and enzyme sensors is the possibility to switch an element "on" or "off" with a light signal. The sensor is operated under UV irradiation which results in the rapid generation of output potential. Visible light restores its original state. The

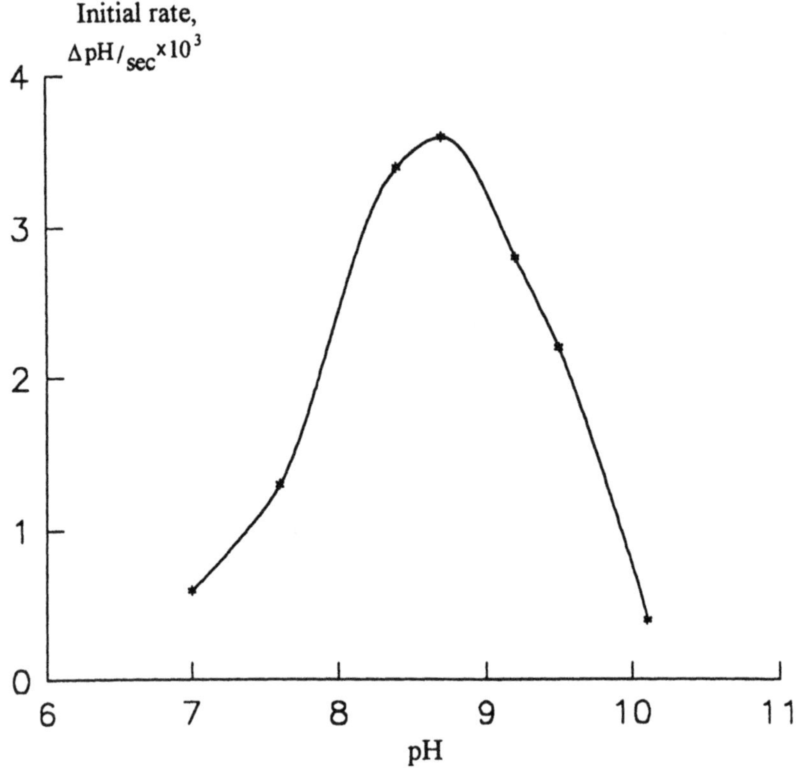

Figure 8. Effect of pH on the response of an IgG sensor. The potentiometric detection of horseradish peroxidase was carried out in 0.15 M. NaCl solution; temperature 20 °C.

potential, in turn, is dependent on the electrolyte composition and presence of substrates. It is necessary to use an appropriate system for photopotential detection. Several experimental devices have been described including a U-shaped two-compartment cell with a membrane separating two electrolyte solutions [31, 32], and a membrane-coated carbon glass electrode [34].

Both a U-shaped cell and electrode are macrosystems about several centimeters in size. We proposed to measure photoinduced potentials across a membrane with a FET. A spiropyran-containing membrane was placed over the gate of FET. Such use of FET is motivated by its basic function of altering output characteristics under the potential applied to the gate. Hence, a miniature phototransducer has been developed by combining a FET with a photosensitive membrane.

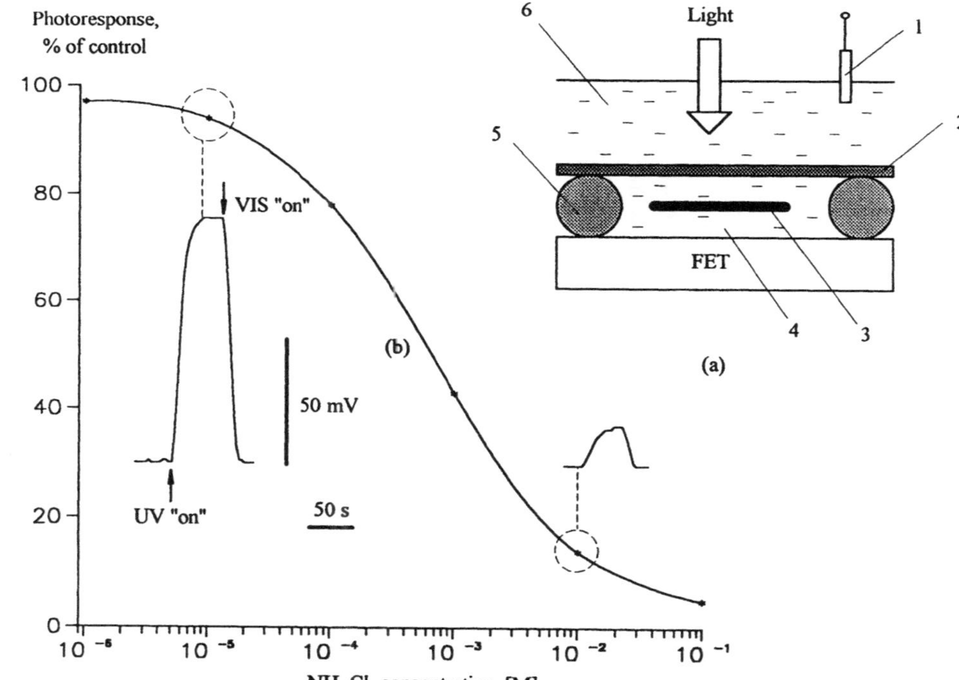

Figure 9. Phototested sensor. (a) schematic representation of the "transistor-membrane" system: (1) reference electrode; (2) spiropyran- and nonactin-modified membrane; (3) light-shielding screen; (4) inner electrolyte; (5) sealing ring; (6) external solution. (b) Dependence of photopotentials generated by the membrane on the concentration of NH_4^+ ions in the medium. Photopotentials are shown which correspond to the low (10^{-5} M) and high (10^{-2} M) concentration of ions. The amplitude of a photoresponse corresponding to the absence of NH_4^+ ions in the medium is taken as 100%.

Membrane Preparation

Polyvinylchloride (PVC) was purchased from Sigma. Dioctyl phtalate (DOP), and tetrahydrofuran (THF) were extra pure reagent grade. A spiropyran compound, 1´-hexadecyl- 3,3´- dimethyl-6-nitrospiro [2H-1-benzopyran-2,2´-indoline], was synthesized according to the procedure reported by Anzai et al.[31]. A membrane with a thickness ca. 0.1 mM was prepared by casting the solution containing 70 mg PVC, 0.15 ml DOP and 5 mg spiropyran in 5 ml THF onto a Petri dish (5.2 cm in diameter) followed by air-drying at room temperature overnight.

Experimental

Schematically, the system "FET + membrane" is shown in the inset of Figure 9. We used light-shielding of the FET to abolish its own photoresponses. A typical photoresponse of the system, "FET + membrane," is shown in the same figure. UV irradiation leads to the generation of potential difference while visible light restores its original state. If the membrane is modified with ionophore, the magnitude of photoresponse is markedly dependent on the concentration of specific ions in the aqueous phase. In our experiments we used the ionophore nonactin which binds NH_4^+ ions. The effect of NH_4 Cl concentration on the photoresponse obtained for our system is presented in Figure 9b. The magnitude of photoresponses was enhanced at lower concentrations of NH_4^+ ions and was suppressed at high concentration.

Thus the system "FET + membrane" is an ammonium-sensitive electrode and can be used to develop biosensors which contain enzymes or microbial cells producing ammonium ions in biochemical reactions. These enzymes can be, for example, urease and asparaginase, together with their respective analyte compounds such as urea and asparagine.

3.5   Metabolic Processes in Expanding Bacterial Populations

Silicon Sensors

In this section, we present results not directly related to biosensor development. It concerns silicon sensors as instruments for research in complex biological systems, which in our case are represented by an expanding population of *E.coli*. We used an FET to detect the pH values at various points within the population during its growth and chemotaxis.

Bacterial chemotaxis is important in a number of microbial processes such as nitrogen fixation, denitrification, pathogenesis of infection and development of biofilms. Figure 10a demonstrates typical wave patterns formed by *E.coli* expanding populations in peptone semisolid agar nutrient medium. The occurrence of first and second waves is due to the chemotactic response and mobility of bacteria. The pH of a nutrient medium is an important parameter which determines the movement of a bacterial population wave. It is well known that pH

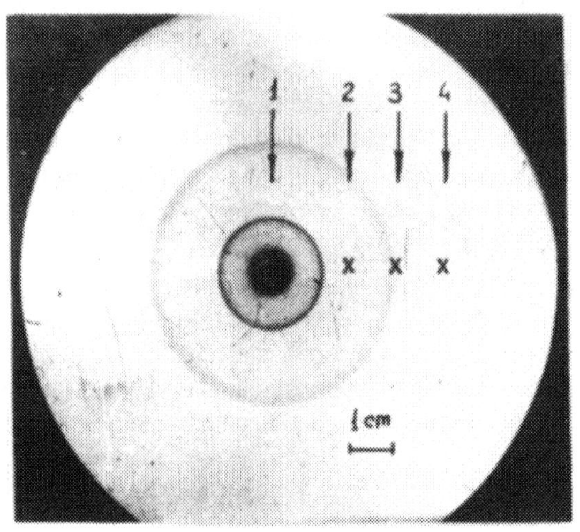

(a)

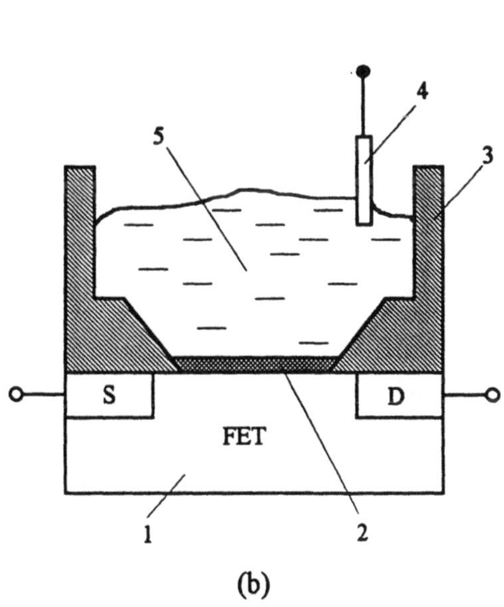

(b)

Figure 10. Typical wave pattern formed by an *E.coli* expanding population in peptone medium (**a**) and a schematic representation of the measuring cell (**b**): (1) FET, (2) gate region, (3) microbath, (4) reference electrode, (5) microsample of agar medium.

affects the wave rate as well as nature of wave interaction [35]. Thus the studies of local pH values in expanding colonies might be of interest from different points of view. Conventional glass pH-electrodes fail to register these changes because of the comparatively large volume (1-2 ml) of a sample. At the same time, miniature sensors like the pH-sensitive FET could be useful for studying pH distributions in expanding populations. FET's can be used in two different ways: the first one is based on measuring pH of microsamples taken at different points in the medium; the other consists of using a single transistor or a set of transistors dipped into the medium and linked with a readout system.

Experimental

We have used both of the above approaches. At first we obtained the pH distribution by measuring the pH of microsamples (a preliminary report has already appeared in ref. [36]). Figure 10b is a schematic representation of an FET-based

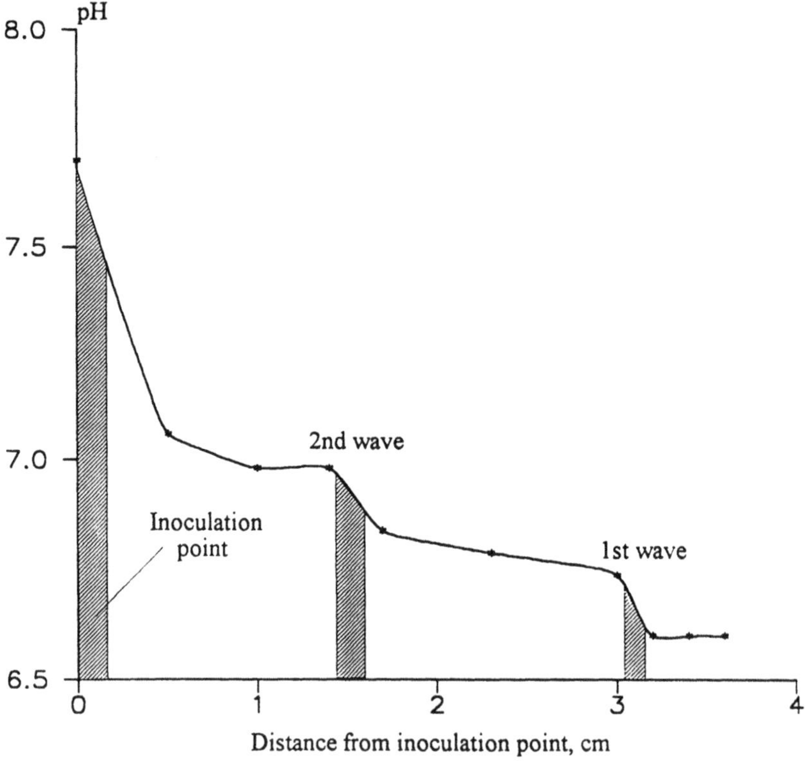

Figure 11. pH profile in the *E.coli* population 4 h after inoculation. The pH changes are most significant in the regions of wave formation and in the inoculation point. The nutrient medium is peptone.

measuring cell with a microbath. The microsamples (10 µl) were taken at the points located: (a) 10-15 mm before the first expanding population wave (control, point "4", Figure 10a); (b) on the wave front(s) point "3" corresponds to the first wave); (c) behind the front (s), point "2"; and (d) at the inoculation point ("4"). The bacterial population waves were formed after inoculation of *E.coli* (strain C) in agar nutrient medium containing peptone (1%, w/v), and NaCl (0,5% w/v). The agar content was 0.31%. The medium was poured into Petri dishes to a depth of 3 mm. Incubation and measurements were at 36 °C.

Figure 11 plots the pH profile of a population growing on peptone. The first and second waves evoked the alkalization of the medium; the most significant pH changes occur in the inoculation point.

In addition to this discrete method of pH registration we also used continuous detection. One transistor was used, though there are no essential restrictions on the number of sensors applied. For instance, multichannel regis-

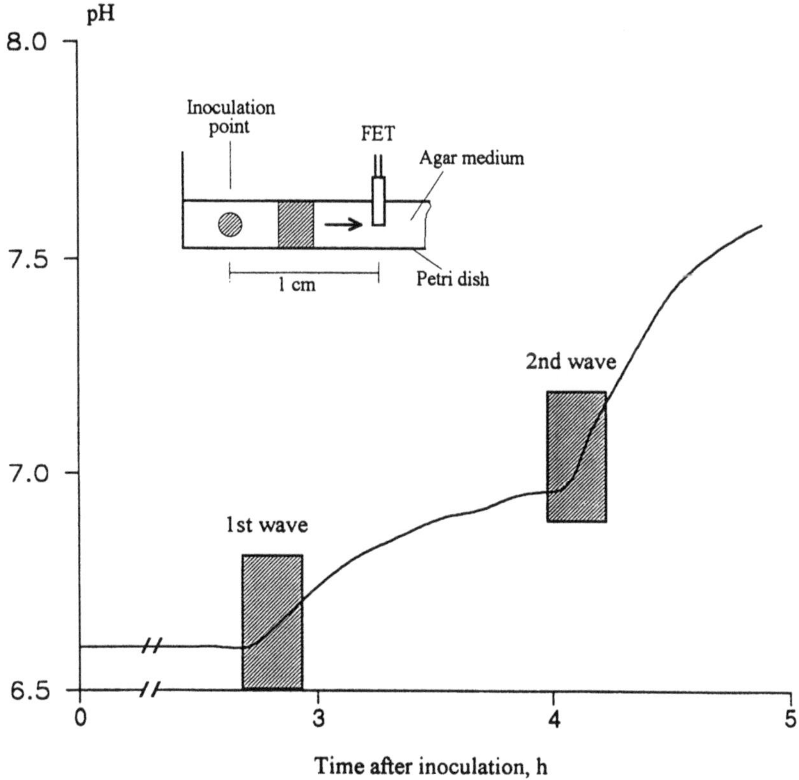

Figure 12. Continuous registration of pH using a FET in the agar medium. The position of the FET relative to the inoculation point is shown in the inset. The first wave evokes alkalization which is enhanced by the second wave.

tration could be made. Figure 12 demonstrates pH changes in the transistor region during the passing of bacterial waves through it. As in the previous example, the nutrient medium was peptone, in which alkalization occurs. The character of the process is in accord with the evaluation made by the discrete method.

We would like to note in conclusion that the techniques developed and information obtained could be useful for general understanding of the metabolism in bacterial colonies, as well as for studies of special issues such as interaction of bacterial wave, and appearance of regular structures [35].

4. CONCLUDING REMARKS

We have developed several types of biosensors. One of them is a sensor for glucose determination. It is still a model system and we are currently making experiments which will allow us to estimate the sensitivity of this sensor in analysis of glucose content in blood. Besides, we plan to create on the basis of this model a sensor for sucrose assay. This can be achieved by use of a combined receptor element, namely, microbial cells and an immobilized enzyme. We have alredy obtained preliminary positive results in this way. We plan to test a sensor for detection of human IgG on samples of human blood plasma, i.e. to check its functioning in real conditions of clinical analyses. In experiments with spirobenzopyran-based phototested sensor we are going to incorporate into membranes enzymes, such as urease and adenosine deaminase whose reaction products are NH_4^+ ions. On this base we might construct sensors for analysis of urea and adenosine, whose determination is vital and urgent in many circumstances. In microbial population investigations, we used a very simple approach to measure the concentration of products of biochemical reactions. Now this method is available and we hope the approach can be informative in different studies of chemotaxis patterns.

ACKNOWLEDGMENTS

The authors thank T.P. Eliseyeva for providing field-effect transistors; M.V. Donova for developing the method of immobilization of microbial cells on transistors; E.M. Gavrilova and A.M. Egorov for consultations and reagents required for studies of the properties of the immunosensor model; M.A.Manenkova and N.F. Kazanskaya for preparation of membranes containing spiropyran and nonactin and discussion of the results; A.B. Medvinsky for assistance in experiments with bacterial populations; Professor A.M. Boronin for attention to this work and useful discussions. Part of this work was supported by a grant from the State Research Program 08.05, "Newest Methods of Bioengineering," and subprogram, "Engineering in Enzymology."

REFERENCES

[1]   Blackburn, G.F.(1987). In: Biosensors. Fundamentals and Applications, pp.481-514. (Turner, A.P.F., Karube, I. and Wilson, G.S., Eds.), Oxford University Press, New York.

[2]   Bergveld, P. (1986). The development and application of FET-based biosensors. *Biosensors, 2*, 15-33.

[3]   Dransfeld, I. Hintsche, R. (1990). Biosensors for glucose and lactate using fluoride ion sensitive field effect transistors. *Anal.Lett.*, **23**, 437-450.

[4]   Brunink, A.J., Haak, R., Bomer, G., Reinhoudt, N., et.al. (1991). Chemically modified field-effect transistors; a sodium ion selective sensor based on calix [4] arene receptor molecules. *Anal. Chim. Acta,* **254**, 75-80.

[5]   Kimura, K., Matsute, K., Yokoyama, M. (1991). Anion-sensitive field effect transistors based on polyion complex bilayers. *Anal.Chim.Acta,* **252**, 41-46.

[6]   Hafeman, D.G., Parce, J.W., McConnell, H.M.(1988). Light-addressable potentiometric sensor for biochemical systems. *Science*, **240**, 1182-1185.

[7]   Parce, J.W., Owicki, J.C., Kercso, K.M., Sigal, G.B., et.al.(1989). Detection of cell-affecting agents with a silicon biosensor. *Science*, **246**, 243-247.

[8]   Bergveld, P., Sibbald, A. (1988). Analytical and biomedical applications of ion-selective field-effect transistors. In: *Comprehensive Analytical Chemistry Series* (Svehla, G., Ed.), vol. 23, p. 188. Elsevier Sciense Publishers, Amsterdam.

[9]   Van Der Schoot, B.H., Bergveld, P. (1987). The pH-static enzyme sensor. An ISFET-based enzyme sensor, insensitive to buffer capacity of the sample. *Anal. Chim. Acta,* **199**, 157-160.

[10] Bergveld, P.(1991). A critical evaluation of direct electrical protein detection methods. *Biosen. Bioelectron.*, **6**, 55-72.

[11] Schasfoort, R.B.M., Bergveld, P.,Bomer, J., Kooyman, R.P.H., et.al. (1989). Modulation of the ISFET response by an immunological reaction. *Sens. Actuators*, **17**, 531-535.

[12] Schasfoort, R.B.M., Bergveld, P., Kooyman, R.P.H., Greve, J. (1990). Possibilities and limitations of direct detection of protein charges by means of an immunological field-effect transistor. *Anal. Chim. Acta*, **238**, 323-329.

[13] Schasfoort, R.B.M., Kooyman, R.P.H., Bergveld P., Greve, J. (1990). A new approach to immunoFET operation. *Biosens. Bioelectron.* **5**, 103-124.

[14] Bergveld, P., Kooyman, R.P.H., Greve, J. (1991). The ion-step-induced response of membrane-coated ISFETs: theoretical description and experimental verification. *Biosen. Bioelectron.*, **6**, 477-489.

[15] McKnabb, S., Rupp, R., Tedesco, J.L. (1989). Measuring contaminating DNA in bioreactor derived monoclonals. *Biotechn.*, **7**, 343-347.

[16] Libby, J.M., Wada, H.G. (1989). Detection of Neisseria meningitidis and Yersinia pestis with a Novel silicon-based sensor. *J. Clin. Microbiol.*, **27**, 1456-1459.

[17] Rogers, K.R., Foleyt, M., Altert, S., Kogat, P., et. al. (1991). Light addressable potentiometric biosensor for the detection of anticholinesterases. *Anal.Lett.*, **24**, 191-198.

[18] Riedel, K., Renneberg, R., Wollenberger, U., Kaiser, G., et. al. (1989). Microbial sensors: fundamentals and application for process control. *J. Chem. Tech. Biotechnol.*, **44**, 85-106.

[19] Gotoh, M., Tamiya, E., Seki, A., Shimizu, I., et.al. (1989). Glucose sensor based on an amorphous silicon ISFET. *Anal. Lett.*, **22**, 309-322.

[20] Hanazato, Y., Shiono, S. (1983). Bioelectrode using two hydrogen ion sensitive field effect transistors and a platinum wire pseudo reference electrode. *Anal. Chem.* **17**, 513-518.

[21] Alva, S., Gupta, S.S, Phadke, R.S., Govil, G. (1991). Glucose oxidase immobilized electrode for potentiometric estimation of glucose. *Biosens.Bioelectron.*, **6**, 663-668.

[22] Racek, J. (1991). A yeast biosensor for glucose determination. *Appl. Microbiol.and Biotechnol.*, **34**, 473-477.

[23] Reshetilov, A.N., Donova, M.V., Koshcheyenko, K.A. (1992). Immobilized cells of *Gluconobacter oxydans* as a receptor element of a glucose sensor. *Applied Biochem. Microbiol.* (Russian), **28**, 518- 524.

[24] Voller, A., Bartlett, A. and Bidwell, D. (1981). *Immunoassays for the 80s*, p. 508. University Park Press, Baltimore.

[25] Glazier, S.A., Rechnitz, G.A. (1991). Preparation and characterization of IgG-coated membranes for possible use as solid phases in enzyme immunosensors. *Anal.Lett.*, **24**, 1347-1362.

[26] Seki, A., Tamiya, E. Karube, I. (1990). Sensitive amplification immunoassay for human IgG and anti-human IgG by using an enzyme cascade system. *Anal. Chim. Acta*, **232**, 267-271.

[27] Nakamura, N., Nashimoto, K, Matsunage T. (1991). Immunoassay method for the determination of immunoglobulin G using bacterial magnetic particles. *Anal. Chim.*, **63**, 268-272.

[28] Sakai, H., Kaneki, N., Hara, H., Ito, K. (1990). Availability and development of an enzyme immunomicrosensor based on an ISFET for human immunoglobulins. *Anal. Chim. Acta*, **230**, 189-193.

[29] Campbell, A.K.(1988). *Chemiluminescense Principles and Applications in Biology and Medicine*, p. 608. VCH, Ellis Horwood.

[30] Anzai, J., Hasebe, Y., Ueno, A. and Osa, T.(1987) Photo-excitable membranes. Photoinduced potential Changes across polyvinyl chloride/spirobenzopyran membranes doped with valinomycin and nonactin. *Chem. Soc. Jpn.*, **60**, 3169-3173.

[31] Anzai, J., Hasebe, Y., Ueno, A., Osa, T. (1987). Photo-excitable membranes. Effects of nonactin as a membrane additive on the photoresponse of polyvinyl chloride. *Bull. Chem. Soc. Jpn.*, **60**, 1515-1516.

[32] Hasebe, Y., Anzai, J., Ueno, A. Osa, T. (1988). Photoinduced potentiometric response of poly(vinyl chloride)/spirobenzopyran/ crown ether composite membranes modified with urease. *J. Phys. Org. Chem.*, **1**, 309-315.

[33] Hasebe, Y., Anzai, J., Ueno, A., Osa, T. (1989). Photo-excitable membranes.Ion-selective photoresponse of polyvinyl chloride. Membranes doped with nonactin and valinomycin. *Chem. Pharm. Bull.*, **37**, 1307-1310.

[34] Chieh, C., Sakai, Y., Hasebe, Y., Anzai, J., et. al. (1989). Photo-switchable ion and enzyme sensors. Photoinduced potentiometric response of glassy carbon electrode coated with polymer or polymer/enzyme dual membrane. *Chem. Pharm. Bull.*, **37**, 3316-3319.

[35] Ivanitsky, G.R., Medvinsky, A.B., Tsyganov, M.A. (1991). From disorder to order as applied to the movement of micro-organisms. *Sov. Phys. Usp.*, **34**, 289-316.

[36] Reshetilov, A.N., Medvinsky, A.B., Eliseeva, T.P., Boronin, A.M., et.al. (1992). pH track of expanding bacterial populations. *FEBS Microbiol.Lett.*, **94**, 59-62.

AMPHIPHILIC POLYELECTROLYTES AS THE BASIS OF THE NEW GENERATION OF BIOSENSORS

I. N. Kurochkin

OUTLINE

Advances in Biosensors
Volume 3, pages 77-109.
Copyright © 1995 by JAI Press Inc.
All rights of reproduction in any form reserved.
ISBN:1-55938-535-9

ABSTRACT

This chapter describes the base technology for the production of LB films of protein using a new type of protective amphiphilic polymers—amphiphilic polyelectrolytes. The present work reports the results of the investigation of the structural parameters of the novel protein LB films by scanning tunneling and atomic force microscopy. Physicochemical investigation of the LB-film formation process and analysis of the functional activities of enzymes and antibodies in these films permit the mechanism of film formatioan to be proposed and suggest real possibilities for these novel films in biosensors.

1. INTRODUCTION

Biomolecules that can be used as specific analytical reagents (or selector elements) such as enzymes, antibodies, DNA and receptors are of particular interest to analytical chemists and biotechnologists. The limiting factor in the analytical applications of such selector elements has been the problem of obtaining them in sufficient quantity and with the necessary activity, long-term stability, and high reproducibility for their usage *in vitro*. Recent technological advances regarding their immobilization on solid supports, thin films, and membranes has overcome such a problem [1]. Various methods of immobilizing a protein and nucleic acids have been reported [1] and widely used in biosensor development.

However, progress and iinteraction between modern information technologies and biotechnologies have recently revealed additional limiting factors in the analytical and technological application of biomolecules. Generally speaking, information is not sufficient to solve the problems of obtaining stable, active and reproducible films or supports with selector elements. Thus it is necessary to solve the problem of structural organization (orientation and coupling different biomolecules, integration of material function) and regulation of the functional activities of the selector elements in such films and supports. This is necessary for obtaining a new level of sensitivity and selectivity in biosensors and is key to the development of bioelectronics, new materials, and nano-technologies.

Recently, the Langmuir–Blodgett (LB) method has been reported as a new biomolecule immobilizing method. The LB method and its modifications has the following advantages [1]:

1. There is little possibility of injuring the activity of the protein.
2. It is possible to produce very thin films (even a few hundred angstroms); thus, we can expect a rapid response from the resulting sensor.
3. The sensitivity and specificity of sensors and functional properties of the proteins in the LB films can be easily controlled by changing the techno-

logical parameters of the LB film preparation process (see results presented in this report).

4. The sensitivity of sensors can be easily controlled by changing the number of transferred films.
5. Since a heterostructure film can be produced, the integration of material functions as well as the orientation and coupling of different biomolecules will be possible.

Langmuir himself is one of the initiators in the study of protein films at the air–water interface [2]. Until recently, however, little was known about the structural aspects of such films and about the connection between structural aspects and functional activities of biomolecules in these films. Denaturation of proteins at the air-water interface and marked decrease in functional activity of these biomolecules has caused difficulty with such investigations and contributed to the lack of structural information on protein monolayers [1-3] and the application of this approach in biosensors. Recent technical advances have now made possible the preparation of enzyme and antibody LB-films with native functional activity and extremely high surface density and even the two-dimensional crystallization of proteins [2,3]. Several approaches for obtaining such protein LB films have been described previously:

- at the air-organic solvent interface [4];
- at the air-water solution with high ionic strength [4,5];
- at the surface of a liquid metal (like mercury) [6];
- at the air-water interface using protective lipid or other surfactant monomolecular layers [7-10];
- at the air-water interface using protective monolayer from cross-linked inert protein (like bovine serum albumin) [11];
- at the air-water interface using protective amphiphilic polymer structures [2].

The purpose of this review is to describe the results of investigations at the Department of Biosensors of the Research Center of Molecular Diagnostics and Therapy (Moscow, Russia) dealing with development of a base technology for the production LB-films of proteins using a new type of protective amphiphilic polymers—amphiphilic polyelectrolytes based on alkylated polyethylenimine (aPEI), alkylated poly-4-vinyl-pyricline (aP4VP), and alkylated copolymers of alkyl-substituted acrylic acid and alkyl-substituted acrylates. All polymers were synthesized in the laboratory of Dr. A.A. Yaroslavov (Chair of Polymer Compounds of the Chemical Department of Moscow State University).

The background of our interest in amphiphilic polyelectrolytes relates to the fact that real conditions for enzyme function in the living cell, and antibodies in the blood usually contain high concentration of polyelectrolytes [12,13]. Polyelectrolytes not only stabilize the protein conformation, but also initiate assembly of multimolecular protein structures and permit regulation of the kinetic properties of enzymes, and receptors, and the affinity of antibodies [12-19].

The present work reports the results of investigating structural parameters of novel protein LB-films, using protective amphiphilic polyelectrolytes, scanning tunneling microscopy (STM), and atomic force microscopy (AFM). Physico-chemical investigation of the LB-film formation process and analysis of the functional activities of enzymes and antibodies in these films permit the mechanism of film formation to be proposed and suggest real possibilities for novel films in biosensors.

Analytical applications using these films have, among others, the following attributes:

1. Low nonspecific adsorption during their utilization in biosensor analysis, immunological and enzymatic analysis, and other analytic applications.
2. Extremely high density of the protein molecules;
3. High density of functional groups, such as carboxylic, hydroxy, amino, and other groups introduced in the polymers for covalent immobilization of protein.
4. Prevention of protein and other selector element inactivation.
5. Regularity in the packing of polymer chains and proteins in the films.
6. Enhanced affinity constant, kinetic constant, and electrochemical proper-ties of the respective proteins and other selector elements.
7. Enabled coupling of different selector elements in the films.

2.   AMPHIPHILIC POLYMERS

LB-films of the present work use the following amphiphilic polymers:

(i)   Poly-4-vinyl-pyridine modified with suitable content of alkyl chains and having a specified length of alkyl chains (aP4VP):

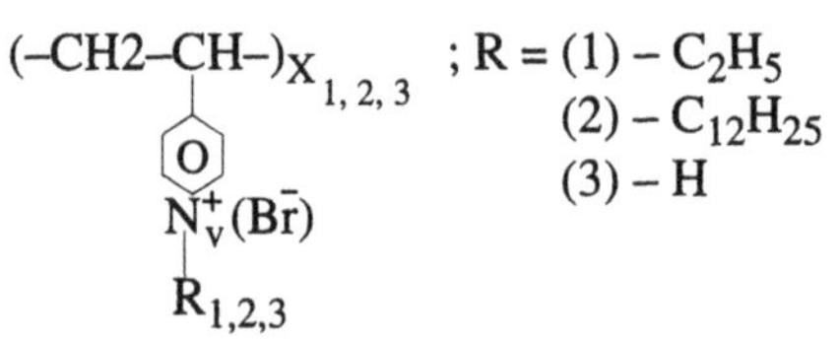

$$(-CH_2-CH-)_{X_{1,2,3}} \quad ; R = (1) - C_2H_5$$
$$(2) - C_{12}H_{25}$$
$$(3) - H$$

with pyridinium ring bearing $N_v^+(Br^-)$ and $R_{1,2,3}$

(ii)   Linear polyethylenimine modified with suitable content and length of alkyl chains (alPEI):

$$(-CH_2-CH_2-NH-)_X \ldots (-CH_2-CH_2-N-)_y$$
$$R = -C_{12}H_{25} \qquad R$$

(iii)   Branched polyethylenimine modified with suitable content and length of alkyl chains (abPEI):

$$(-CH_2-CH_2-NH-) \ldots (-CH_2-CH_2-N-) \cdot (-CH_2-CH_2-N-)$$

$$R_1 = -C_{12}H_{25}; \; R_2 = -C_{12}H_{25} \quad ; \qquad (CH_2)_2 \qquad\qquad R_1$$
$$-(CH_2-NH-R_3) \qquad NH-R_2$$
$$R_3 = -C_{12}H_{25}, -H;$$

(iv)   Copolymers of alkyl-substituted acrylic acids and alkyl-substituted acrylates having specified alkyl chain length and in suitable proportions:

$$\begin{array}{ccc} & R_1 & \\ & | & \\ (-CH_2-C-)_x \ldots (-CH_2-C-)_y & ; & R_1 = -CH_3 \\ | \qquad\qquad | & & R_2 = -C_{12}H_{25} \\ C=O \qquad\qquad C=O & & \\ | \qquad\qquad | & & \\ OH \qquad\qquad O=R_2 & & \end{array}$$

The actual proportion of different monomeric units will be indicated in each experiment.

p-A isotherms (where, p is the surface pressure and A is area per one monomeric unit or one hydrophobic radical) for some amphiphilic polyelectrolytes like: aP4VP with 70% of monomeric units containing cetyl radical (aP4VP-70); alPEI with 35% (alPEI-35) and 70% (alPEI-70) of monomeric units containing cetyl radical; abPEI with 12% (abPEI-12) and 70% (abPEI-70) of monomeric units containing cetyl radical; copolymers of metacrylic acid and lauryl methacrylate with 25% (cLMMA-25), 40% (cLMMA-40) and 90% (cLMMA-90) lauryl methacrylate monomeric units, were obtained (Figure 1) using standard Langmuir technique. The data are normalized form normalizing per one cetyl radical containing monomeric unit (Figure 1).

Table 1 shows the area per one monomeric unit containing cetyl radicals and area per one monomeric unit of polymer chains. The best surface density of polymer chains in the LB films was seen for abPEI-12 and aP4VP-70.

High density of the polymer chains of abPEI-12 and aP4VP-70 in the monomolecular films obtained by searching the p-A isotherm has very good correlation with atomic force microscope (AFM) and scanning tunneling microscope (STM) pictures after transferring these films to the highly oriented pyrolitic graphite surface (Figures 2, 3). AFM and STM investigation of the LB films from abPEI-12 was performed by A. Pavelev, L. J. Pelliniemi, and J. Maki at Turku University and supported by the Finnish-Russian Biotechnology Laboratory (JBL), (Prof. Timo Korpela, director). Good regularity of the polymer chains in this film is also demonstrated by Figures 2 and 3.

Thus, the results obtained indicate good amphiphilic properties and good polymer chain density in films of the new type of amphiphilic polyelectrolytes.

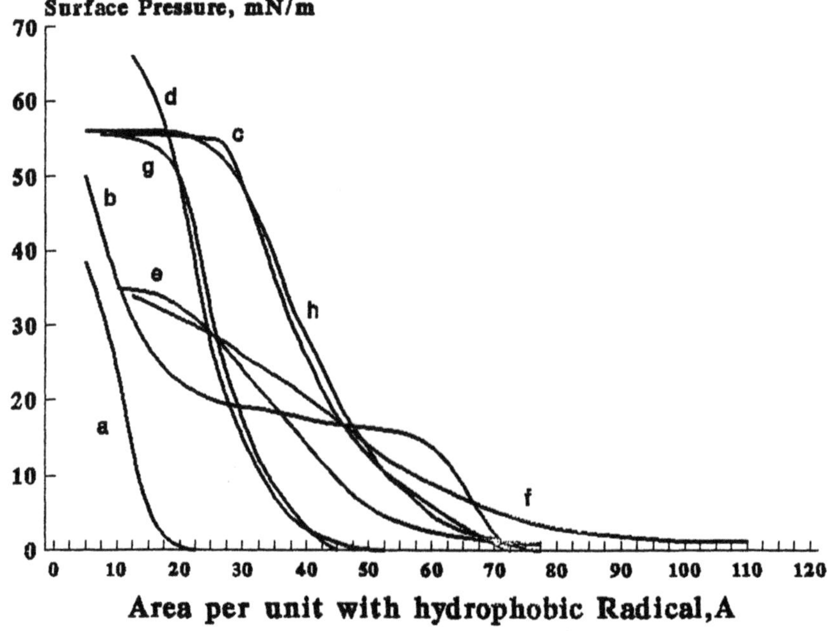

Figure 1. "p-A" isotherms of the amphiphilic polyelectrolytes. (**a**) aP4VP-70, (**b**) cLMMA-90, (**c**) alPEI-35, (**d**) abPEI-12, (**e**) cLMMA-40, (**f**) cLMMA-25, (**d**) alPEI-70, (h) abPEI-70.

Table 1. Area Per Different Parts of Amphiphilic Polymers in the LB Films

Polymer Type	Area Per Monomeric Unit Containing Cetyl Radical, A^2	Area per Common Monomeric Unit, A^2
abPEI-12	31.0	6.2
alPEI-35	50.0	14.9
alPEI-70	32.0	23.4
abPEI-70	55.0	37.9
aP4VP-70	17.0	11.4
cLMMA-25	76.0	19.6
cLMMA-40	56.0	19.2
cLMMA-90	28.0	20.0

2. PROTEIN MONOMOLECULAR FILMS BASED ON AMPHIPHILIC POLYELECTROLYTES

The of LB process includes the following steps:

1. A protein (antibody, enzymes, insulin) solution in 5-10 mM tris (or phosphate) buffer, pH 8.2 (6.5), 10 ug/ml, was placed in a Langmuir device.

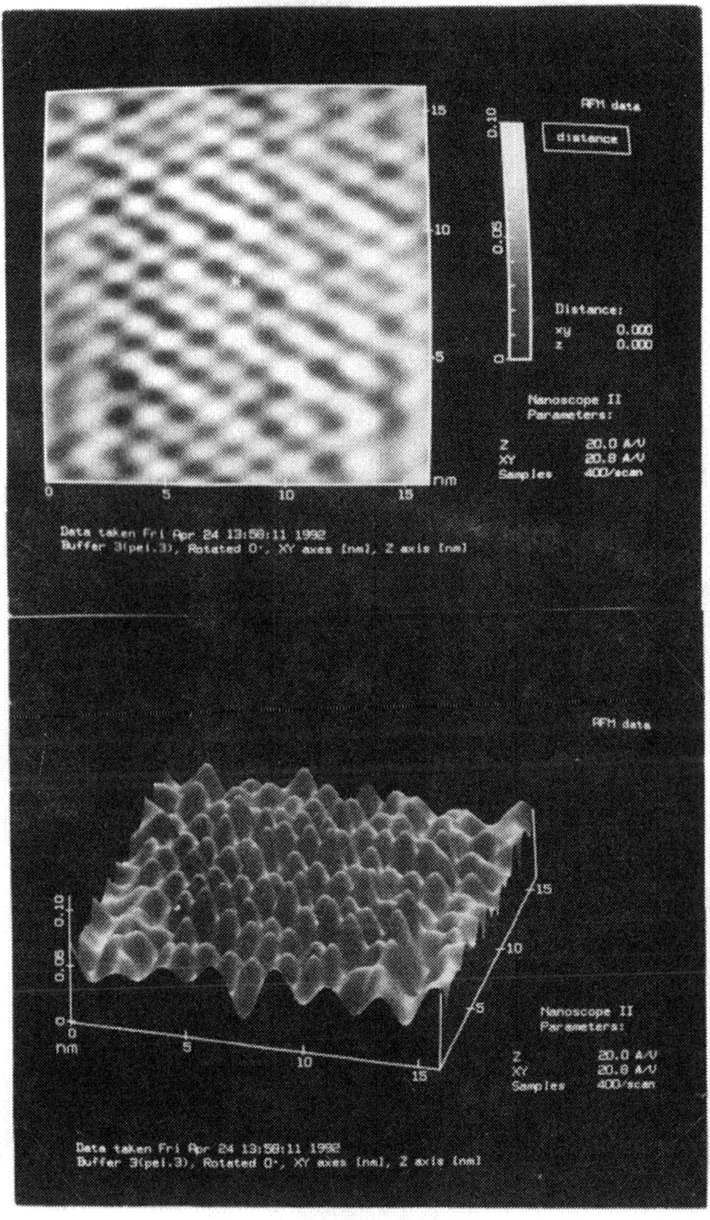

Figure 2. AFM image of the LB films of abPEI-12. (**a**) two-dimensional image, (**b**) three-dimensional reconstruction.

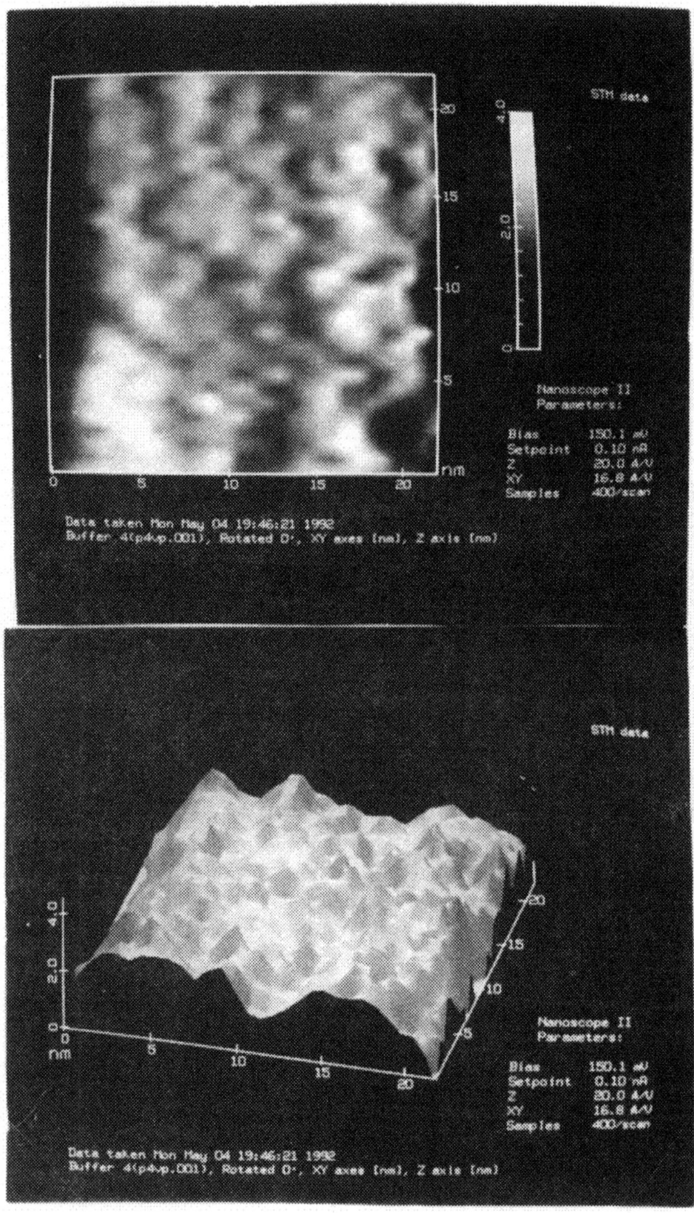

Figure 3. STM image of the LB films of aP4VP-70. (**a**) two-dimensional image, (**b**) three-dimensional reconstruction.

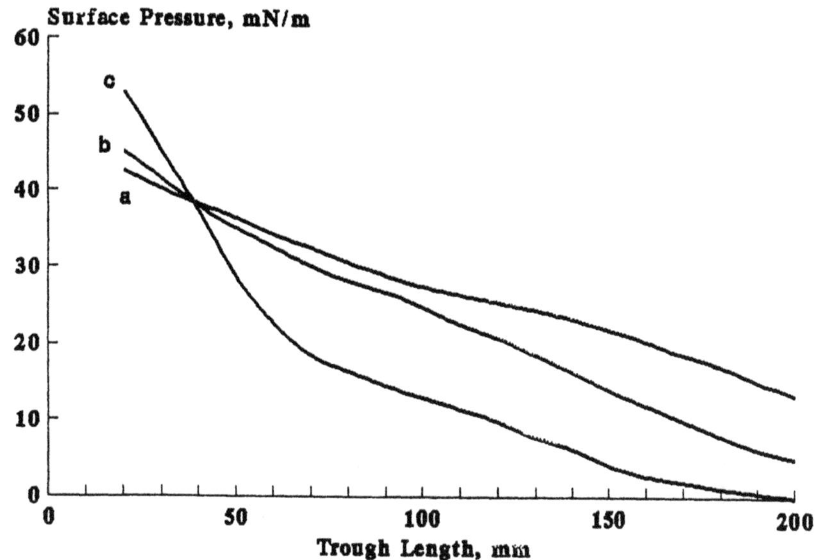

Figure 4. "p-A" isotherms for HB$_S$-antibody - abPEI-12 LB films. (**a**) abPEI-12 with HBS-antibodies, (**b**) pure antibodies, (**c**) - pure abPEI-12.

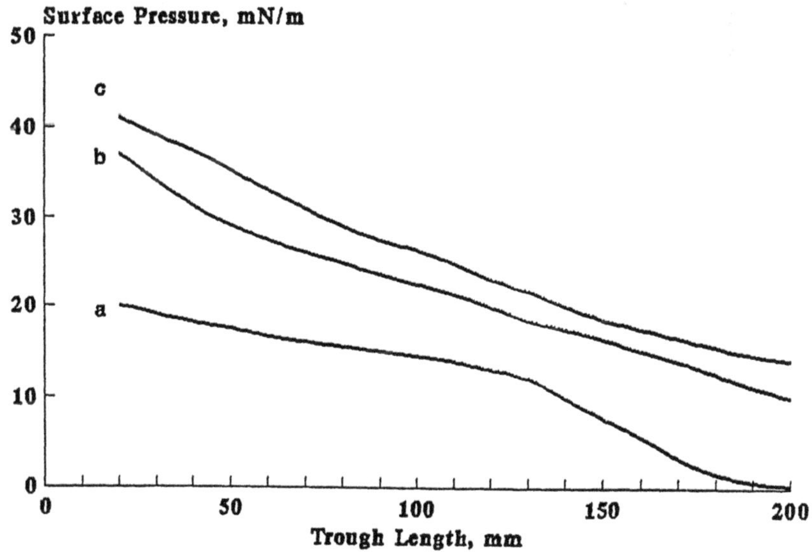

Figure 5. "p-A" isotherms for HB$_S$-antibody - cLMMA-90 LB films. (**a**) pure cLMMA-90, (**b**) pure HBS-antibodies, (**c**) - cLMMA-90 with HBS-antibodies.

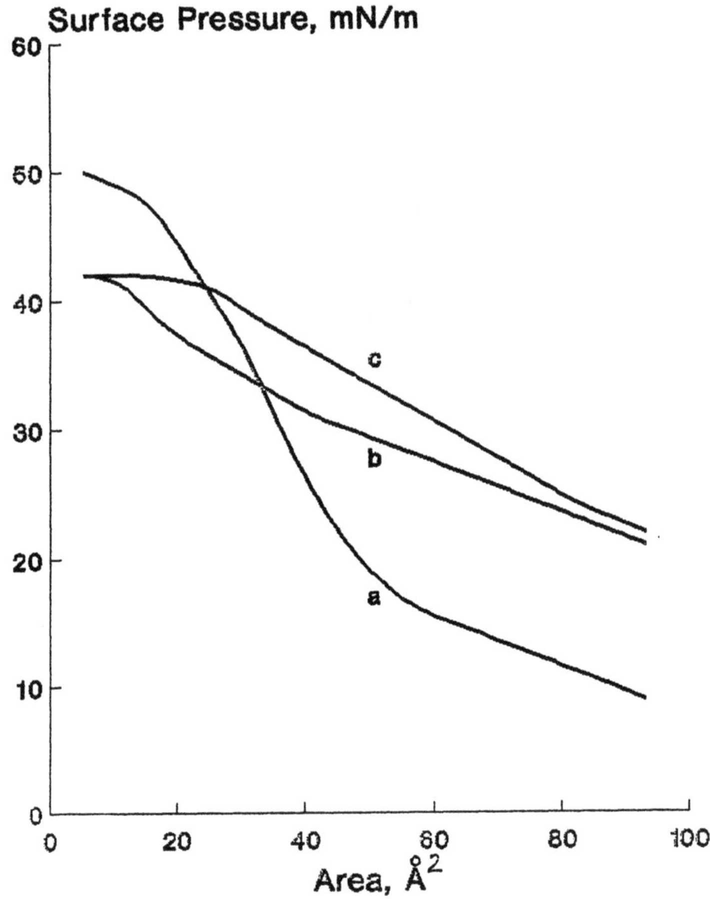

Figure 6. "p-A" isotherms for MAO - abPEI-12 LB films. (**a**) pure abPEI-12, (**b**) pure MAO, (**c**) abPEI-12 with MAO.

2. After 5 minutes, the surface was cleaned and a solution of suitable amphiphilic polyelectrolyte in chloroform (concentration 1 mg/ml) was added until surface pressure reached 5 mN/m. Immediately after this, the surface pressure started to increase.

3. After 15 minutes p-A isotherms were obtained or monomolecular film was transferred onto a suitable surface after pressurizing to 20-40 mN/m using the moving barrier of the Langmuir device. The film can be transferred onto the solid surface by the Langmuir-Schaefer or Langmuir-Blodgett technique. The surface with transferred monolayer is then purged with an air stream at pressure of 0.1-1.0 atm. to remove liquid drops.

p-A isotherms for pure polymer films, for protein-polymer films and for pure protein films were obtained and are illustrated in Figures 4, 5, and 6 for HB_S antibodies with abPEI-12, cLMMA-90, and monoamine oxidase with abPEI-12, respectively. Significant changes of the p-A isotherms due to the strong complex formations on the surface of water subphase were seen. These kind of p-A isotherms permit the following mechanism of protein-polymer interactions during LB-film formation to be proposed. At the first step, protein molecules interact with polymer chains of polyelectrolytes in the two-dimensional gas state on the water surface. During this step, polymer, protein, and protein-polymer complex molecules are on the water surface. For this reason surface pressure was higher than usual. At the second step, due to pressure from the moving barrier pure protein, molecules are translocated from the water surface to the bulk solution—only polymer molecules and polymer-protein complexes keep their position on the water surface. However, part of the polymer chain of the amphiphilic polyelectrolyte produces a strong interaction with protein and for this reason are in solution near the surface. Total surface pressure in this case is lower than the sum of partial surface pressure (Figure 7).

Thus, the strong molecular interactions between amphiphilic polyelectrolytes and protein during the process of the LB-film formation on the surface of the water subphase was seen.

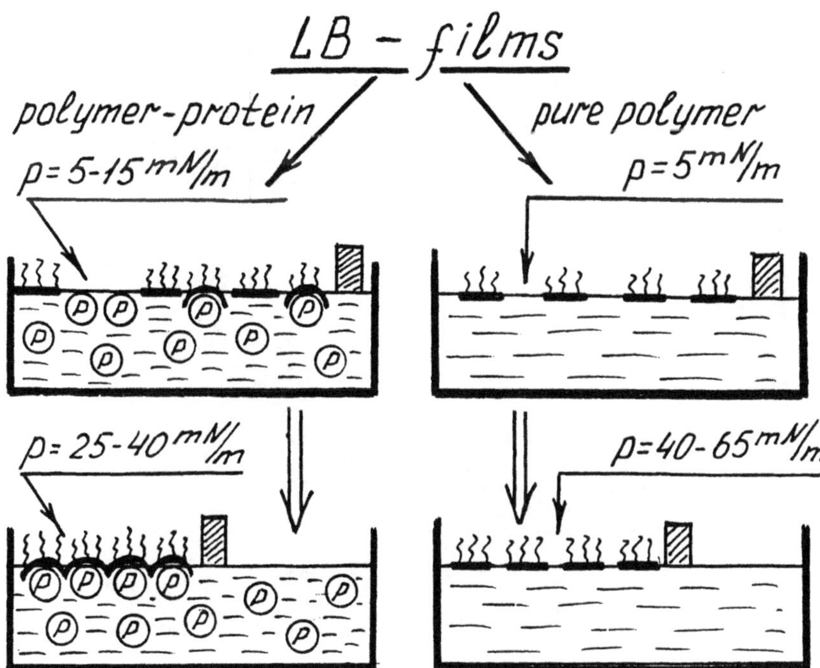

Figure 7. Molecular model of the protein-amphiphilic polyelectrolyte interaction on the surface of water.

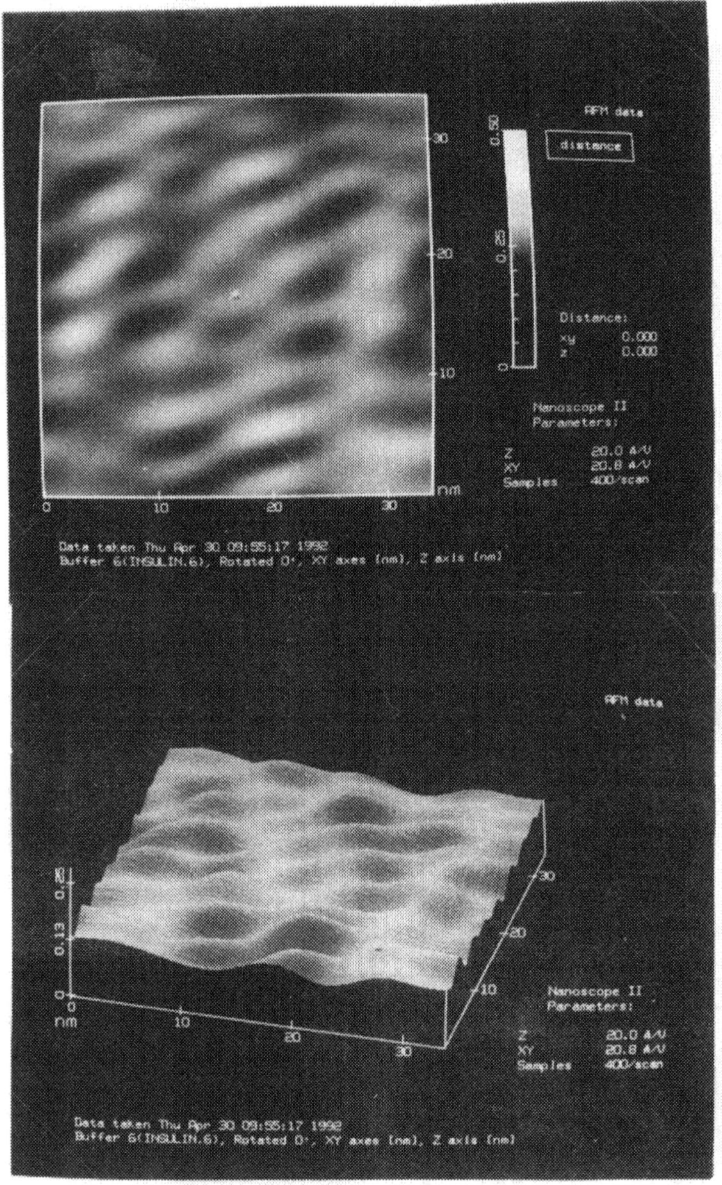

Figure 8. AFM image of the insulin - abPEI-12 LB films. (**a**) two-dimensional image, (**b**) three-dimensional reconstruction.

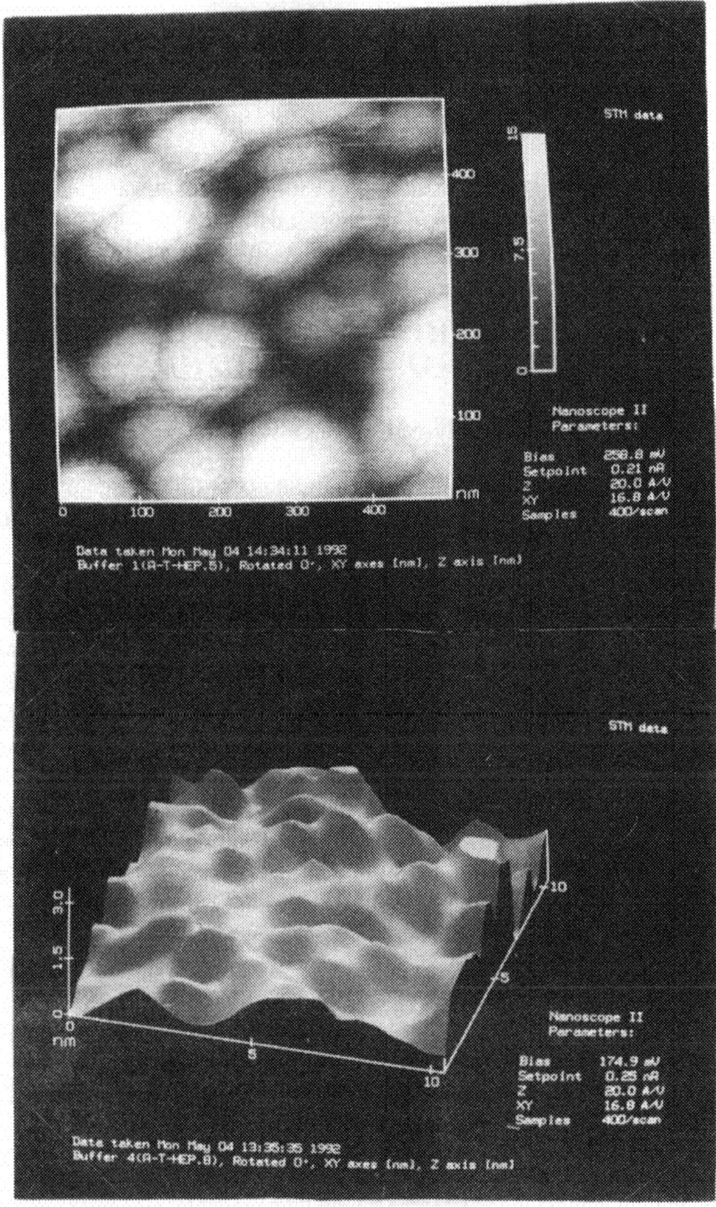

Figure 9. STM image of the HB$_S$-antibodies - cLMMA-90 LB films. (**a**) two-dimensional image, (**b**) three-dimensional reconstruction.

Table 2.   Physicohemical Parameters of the HB_S-Antibody Films

Type of Films	Radical Density of the Ab: Number of Molecules/m^2	Affinity Constant: nM	% of the Nonspecific Binding on the Linear Part of the Isotherm
Pure HB_S antibodies	1.2 x 1015	5.0	51.0
HB_SAb, cLMMA90	7.0 x 1015	0.2	16.2
HB_SAb, alPEI70	3.2 x 1015	0.4	10.9

High surface density and good ordering of protein molecules in the LB films with amphiphilic polyelectrolytes are demonstrated by AFM and STM pictures for insulin and HB_S-antibody monolayers on the surface of HOPG (Figures 8, 9).

4. FUNCTIONAL AND SENSOR PROPERTIES OF PROTEIN LB FILMS

4.1 HB_S-Antibody LB Films

Physicochemical parameters of the HB_S-antibody LB films based on cLMMA-90 and alPEI-70 were calculated from binding isotherms of HB_S-antigen absorption on the films. HB_S-antigen concentration on the HB_S-antibody films were determined by traditional "sandwich" immunoassays using second antibodies labeled by peroxidase. Surface density and affinity constant of the HB_S-antibodies in the films are indicated in Table 2. Table 2 shows that the best amphiphilic polymer for antibody film formation is the cLMMA-90, having a low affinity constant and a high surface density of the antibodies in the film. On the other hand, nonspecific absorption was lower for alPEI-70 application. These data also demonstrate that a polystyrene surface coated with antibody-amphiphilic polymer films may be advantageously used in immunoenzyme analysis with the traditional calorimetric assay. These films increase the sensitivity of the immunological analysis due to the high surface density and good affinity constant of the antibodies in the films.

Coating the antibody-amphiphilic polymer films onto a gold or carbon electrode surface opens good possibilities for the creation of very simple and small immunosensor potentiometric elements. A laboratory prototype of this potentiometric immunosensor has been developed by Dr. A. Eremenko, S. Chernov and A. Ghindilis in our department. The main analytical parameters are as follows: analyte-HB_S-antigen; sensitivity-0.1 ng/ml; stability—3-6 months; response time—2 min; assay time—20-30 min. Two types of molecular constructions for such immunosensors are indicated in Figure 10.

4.2   Catalytic Properties of Glucose Oxidase LB Films

This part of the work was done together with A. Barmin, A. Eremenko and S. Chernov.

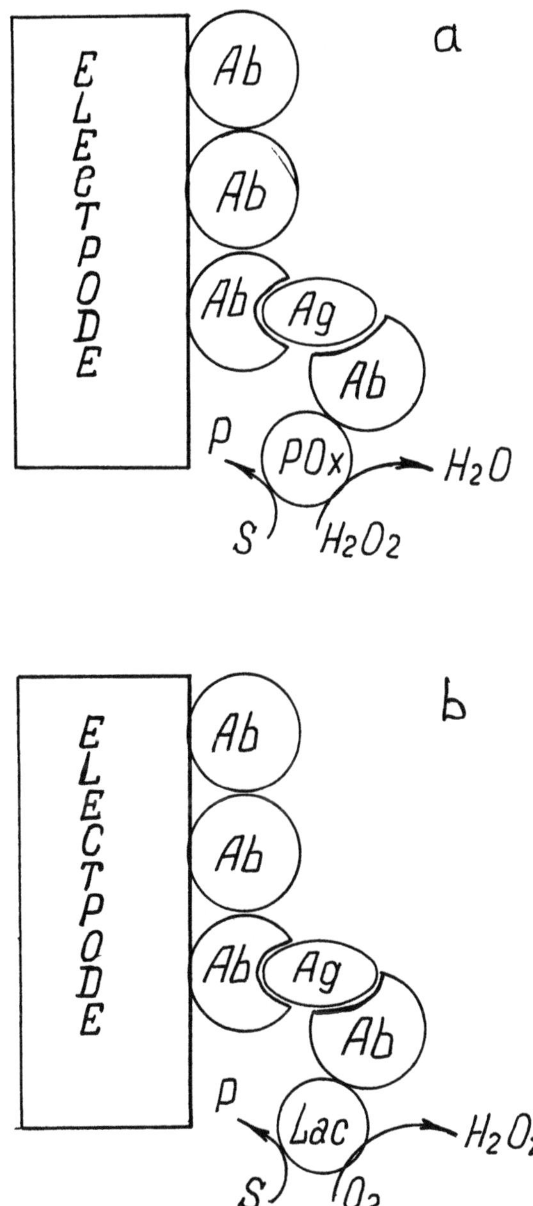

Figure 10. Molecular model of peroxidase (**a**) and laccase (**b**) potentiometric immunosensors.

Formation of glucose oxidase monomolecular films was carried out using the Langmuir-Schaefer method with amphiphilic polymers as described above. When cLMMA-90 was used, the process of film formation was carried out in 10 mM of potassium-sodium phosphate buffer, pH of 5.0, with a surface pressure of 20 mN/m. When abPEI-12 was used, the process was carried out in 10 mM tris-buffer, pH of 8.2, with a surface pressure of 40 mN/m.

A hermetically sealed cell was used in this part of the work. The films were applied to polypropylene membranes selectively permeable to oxygen and tightly fitting the platinum surface of a Clark electrode. Glucose oxidase activity was measured with the Clark electrode on the surface of which a membrane with an enzyme monolayer was fixed. The initial rate of oxygen loss was calculated by the maximal slope of the tangent to the integral kinetic curve of oxygen reduction. Measurement were made in a 1.4 ml hermetically sealed cell including a reaction cell equipped with a magnetic stirrer (rotation rate—250 rw/min) and inlet for glucose solution. The measurements were made at 25 °C in 10 mM phosphate buffer, pH 6.5, which corresponds to an experimentally confirmed optimum of enzyme activity. Oxygenation of the solution according to Henry's coefficient (773 atm/mole·kg H_2O) was 260 µM. The values presented are arithmetic average means from no less than three measurements. The parameters were estimated on a computer using STATGRAPHICS and ENZFITTER programs.

To elucidate the role of the amphiphilic polymers of the present invention in the enzymatic activity of monomolecular glucose films obtained by the Langmuir-Schaefer technique, the activity of membranes modified with monomolecular films of amphiphilic polyelectrolytes (cLMMA-90 or abPEL-12) and glucose oxidase was compared with the activity of membranes modified with monomolecular films of glucose oxidase (glucose 10 mM) only. The data are shown in Figure 11.

The data indicate that membranes covered with a monomolecular glucose oxidase film without amphiphilic polymers are practically inactive. Glucose oxidase films formed on the basis of cLMMA-90 show a relatively low activity. Optimal results were obtained using abPEI-12. Glucose oxidase films obtained with this matrix were taken for analysis of the kinetic characteristics of the immobilized glucose oxidase.

To determine a maximal density of the enzymes on the surface of the polypropylene membranes a comparative analysis of the activities of single-layered and multilayered glucose oxidase amphiphilic polymer films was carried out (glucose 20 mM) (Figure 12).

As the data obtained shows three-layered films have 1.5 times higher activity than single-layered ones. A larger number of layers did not significantly increase the enzymatic activity. This indicates that with three consecutive applications the membrane is packed with the active form of glucose oxidase to a maximal degree. In further experiments four-layered membranes were used as optimal.

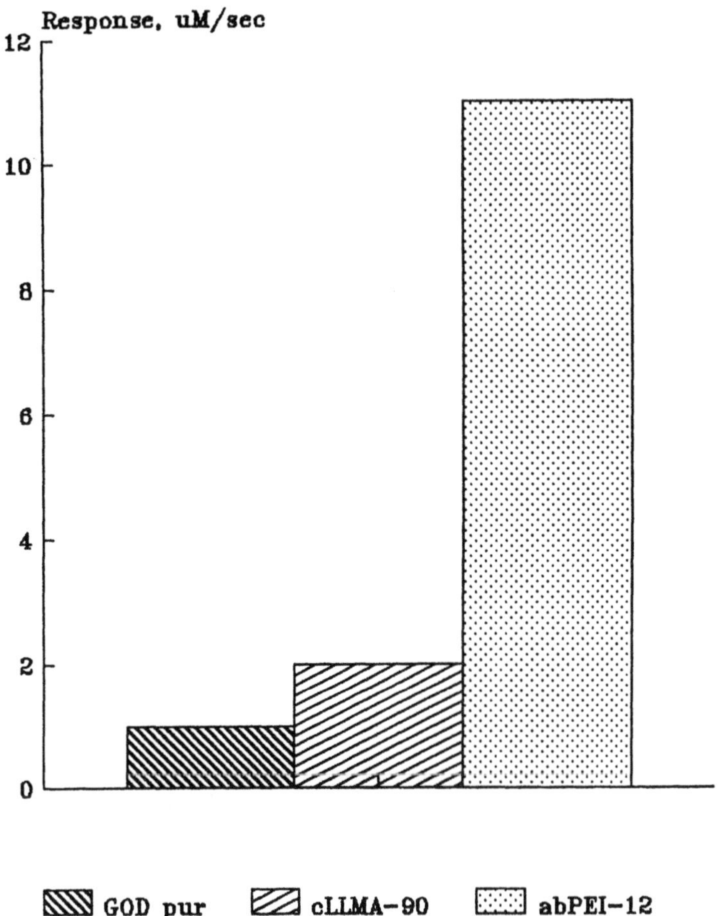

Figure 11. Activity of glucose oxidase (V_0) in LB films: pure GOD (GOD pur) with cLMMA-90 and with abPEI-12 (glucose 10 mM).

Table 3 provides values of kinetic parameters of glucose oxidase in solution and in polymer-enzyme films. As the data obtained indicate, the formation of monomolecular glucose oxidase films results in a significant decrease of the kinetic constant of the enzyme.

This fact is in contrast to numerous data on immobilization of glucose oxidase by traditional methods of immobilization of the enzyme in gel, covalent linkage with glutaraldehyde, and adsorption when an increase of the constants or the preservation of their initial levels is usually observed [20-22]. Thus, the technique used by us for applying the enzyme on a solid surface makes it possible to improve its characteristics by increasing its affinity to substrates.

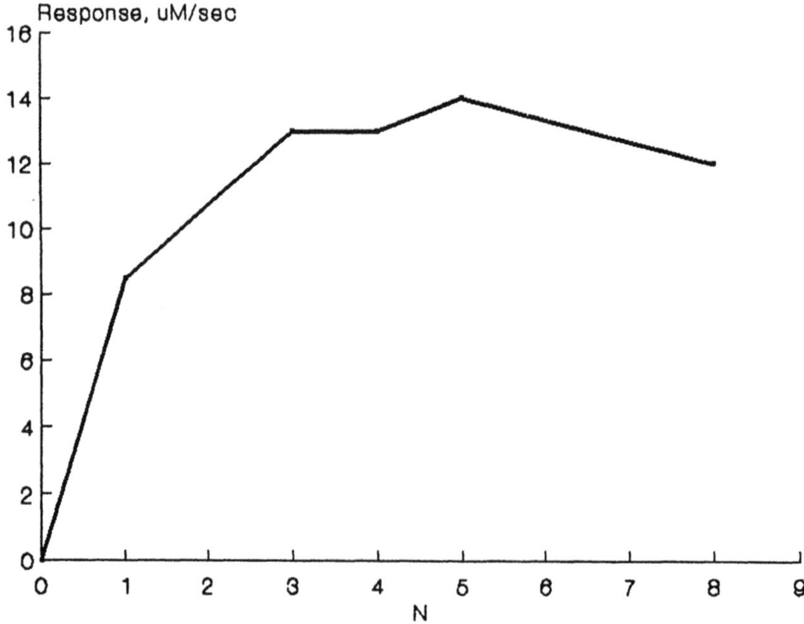

Figure 12. Enzymatic activity of GOD -abPEI-12 LB films depending on the number of monomolecular layers (glucose 20 mM).

Table 3. Kinetic Constants for Dissolved and Immobilized Glucose Oxidase Monomolecular Films

Kinetic Constants	Dissolved Glucose Oxidase	Immobilized Glucose Oxidase
K_{Gl} (mmole/1	30.5 ± 1	8.2 ± 0,3
K_{OX} (μmole/1)	350 ± 40	7.8 ± 1,5
Literature data:		
K_{Gl} (mmole/1)	33 - 110	14
K_{OX} (μmole/1)	200	540

Notes: K_{Gl}-true Michaelis constant for glucose,
K_{OX}-true Michaflis constant for oxygen

Estimation of the analytical possibilities of the electrode with a multilayered monomolecular film of glucose oxidase formed on its surface showed that a minimum glucose concentration under optimal conditions of 1 mM of a final concentration in a cell, i.e. the sensitivity of such electrode is on a par with that of earlier constructed models [23]. A linear relationship for dissolved and immobilized glucose oxidase was observed over a concentration range of 1-10 mM (Figure 13).

The reproducibility of data for dissolved and immobilized glucose oxidase was ±5% and ±15%, respectively. The films withstood without any loss in their

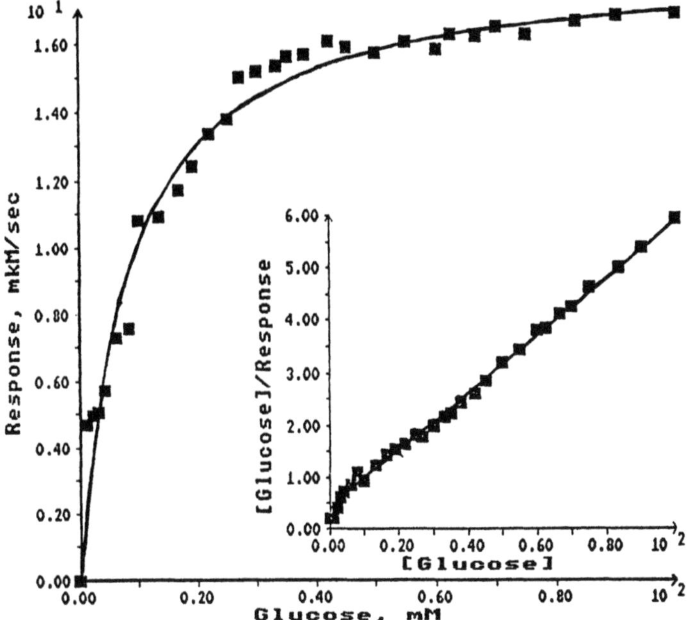

Figure 13. Relationship between the initial rates (V_o) of the enzymatic reaction catalyzed by glucose oxidase in GOD-abPEI-12 LB films and the glucose concentration.

activity no less than 100 repeated measurements of glucose concentration. The activity of immobilized glucose oxidase membranes was 100% after a 2 month storage in the air at room temperature and 50% after 3 months. At the same time, upon constant storage of membranes in a buffer solution, the activity reduced by 50% after 1 week.

The efficiency and simplicity of the employed method of immobilization demonstrates in principle its suitability in the construction of glucose biosensors.

4.3. Ferrocene Derivatives (FD) LB Films and Their Coupling with Glucose Oxidase (GOD)

This part of the work has been done together with A. Barmin and Dr. A. Eremenko and describes the utility of amphiphilic polyelectrolytes to support ferrocene mediators in connection with biosensing of glucose. The mediator properties of FD in LB films obtained on the base of amphiphilic polymers in the absence and presence of GOD in solution and multilayer LB films obtained by electrostatic coadsorption of FD and GOD to amphiphilic polymer were studied in this part of the work.

Cyclic voltammograms of synthesized FD were examined. Figure 14 and Table 4 display different the half-wave potentials ($E_{1/2}$) for equimolar solutions of FD. The maximum catalytic current was obtained for ferrocenecarboxylic acid (FCA) [Figure 14 (A)] in the absence (a) and presence (b) of GOD and glucose in buffer solution. The following trend in catalytic possibilities is observed: FCA > ethyl FCA (EFCA) > butyl FCA (BFCA). GOD, glucose, and buffer solution did not increase the current at $E_{1/2}$ potentials of FD.

The coefficient calculated as $K = I_e/I_o$, where I_e is the current in the presence of enzyme and substrate and I_o is the current without enzyme and substrate at $E_{1/2}$, was used for the estimation of the contribution of the enzyme reaction. Table 4 shows not only different $E_{1/2}$ for FD's, but the different catalytic possibilities of FD's during the enzyme reaction. The best catalytic (mediator) properties were found for ferrocenecarboxylic acid.

Figure 15 depicts p-A (surface pressure-area per one molecule) isotherms of FCA. The pure FCA forms poor monolayer films on the water surface (a). A negligible amperometric response was observed when we studied monomolecular films of FCA transferred onto the graphite electrode surface (data not shown).

The FCA molecule has a negatively-charged carboxylic group and can interact with positively charged amphiphilic polymer groups under certain conditions. The physicochemical properties of FCA monomolecular films obtained on polymeric matrices (derivatives of PEI and PVP) were studied. Cyclic voltammograms of such films after transferring them onto a graphite electrode surface are shown in Figure 16. The ferrocene peak at 285 mV was observed in all cases when the FCA monolayers obtained with the help of abPEI-35 (A), abPEI-12 (B), alPEI-70 (C), abPEI-70 (D), and aP4VP-35 (E). But, the increase of the analytical response after adding glucose was different and was dependent on the polymeric matrices. Table 5 and Figure 16 show that the maximum response in the presence of GOD [Figure 16, A(b)] was observed when abPEI-35 was used as the polymer matrix (K = 2.8, Table 5). Probably, this hydrophobised polymer has the better combination of positively-charged groups and hydrophobic properties for FCA molecule orientation into the LB film.

To examine the mediator properties of EFCA and BFCA after their immobilization on the surface of graphite electrodes, $K = I_e/I_o$ for these derivatives of FCA was estimated. Table 6 shows that the K values for EFCA and BFCA were lower then for FCA.

Table 4. Half-Wave Potentials ($E_{1/2}$) and Current Density for 200 µM Ferrocene Derivatives in Solution

Ferrocene Derivatives	*E1/2 (mV)*	*Current Density at $E_{1/2}$ (nA / mm^2)*		*K = I_e/I_o*
		Base Level	*+ 100 mM Glucose*	
FCA	340	1750	4875	2.8
EFCA	290	850	1750	2.1
BFCA	280	780	1405	1.8

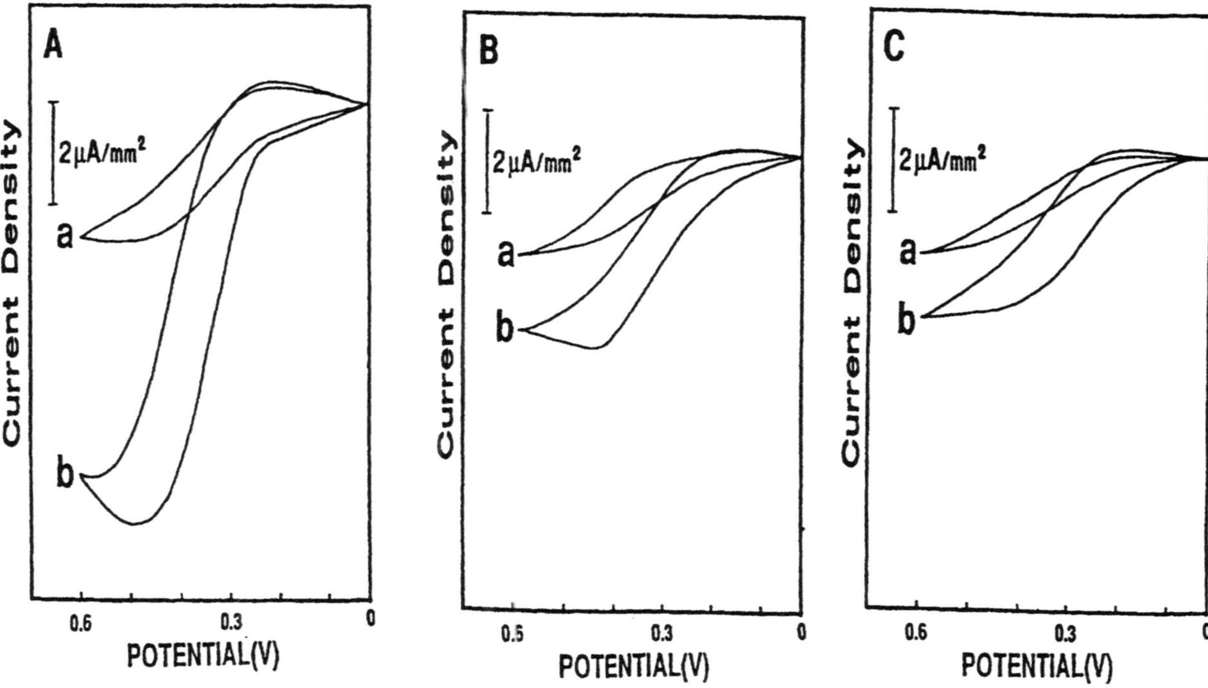

Figure 14. Cyclic voltammograms for 0.2 mM FCA (**A**), EFCA (**B**), and BFCA (**C**) in the absence (**a**) and the presence (**b**) of 10 µM GOD and 0.1 M glucose at graphite electrodes. Scane rate, 5 mV/s; supporting electrolyte 10 mM phosphate buffer (pH 6.5).

Polymers	Current Density at $E_{1/2}(nm/mm^2)$		$K = I_e/I_o$
	Base Level	+ 100mM Glucose	
alPEI-35	554 ± 79	1527 ± 92	2.8
alPEI-12	427 ± 55	1083 ± 162	2.5
alPEI-70	251 ± 78	612 ± 67	2.3
abPEI-70	423 ± 136	940 ± 113	2.2
aP4VP-35	387 ± 147	808 ± 142	2.1

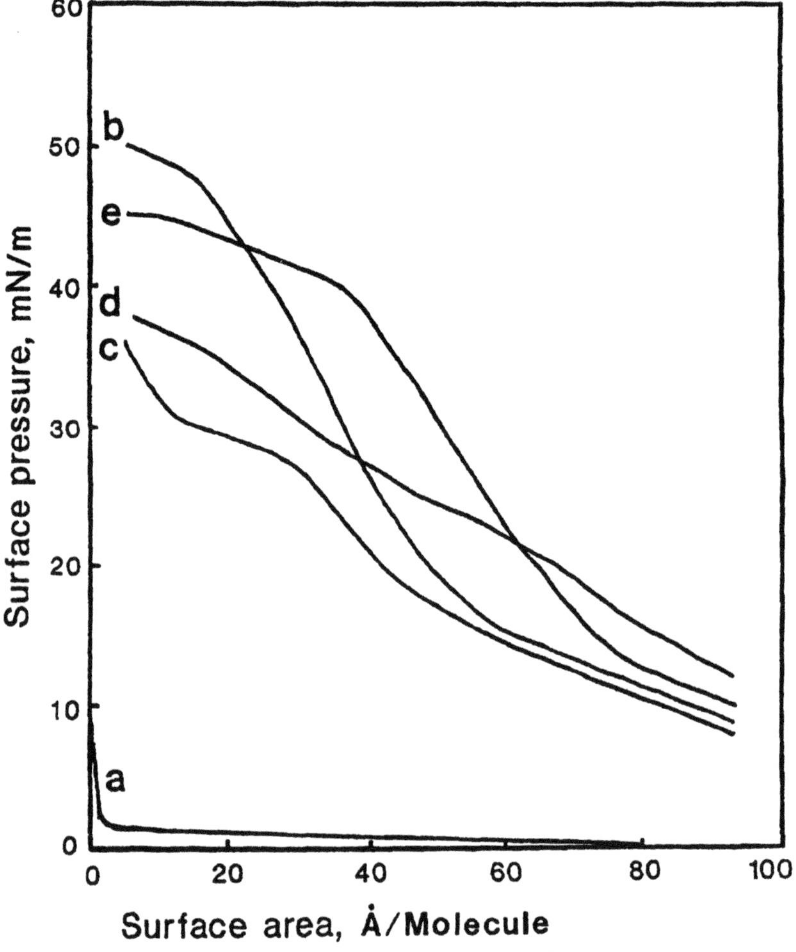

Figure 15. "p-A" isotherms of the LB films for FCA (**a**), abPEI-12 (**b**), FCA preliminary mixed with abPEI-12 (**c**), FCA preliminary mixed with abPEI-12 with GOD (**d**), and abPEI-12 with FCA (**e**) in 10 mM phosphate buffer (pH 8.2).

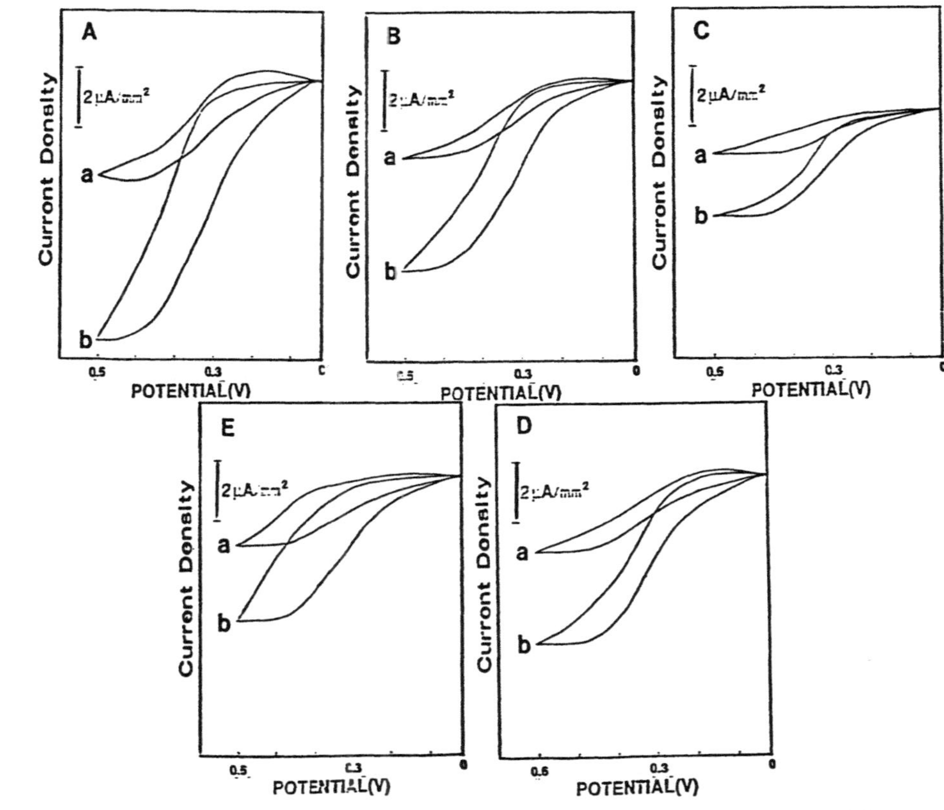

Figure 16. Cyclic voltammograms for FCA-polymer LB films after their coating to graphite electrode for different polymers in the absence (**a**) and presence (**b**) 0.1 M glucose. (**A**) alPEI-35, (**B**) abPEI-12, (**C**) alPEI-70, (**D**) abPEI-70, (**E**) aP4VP-35. Other conditions, as in Figure 14.

Table 6. Current Density at $E_{1/2}$ for Different FD
Immobilized onto graphite Electrodes Using a1PEI-35

Ferrocene Derivatives	*Current Density at $E_{1/2}(na/mm^2)$*		$K = I_e/I_o$
	Base Level	*+ 100mM Glucose*	
FCA	554 ± 79	1527 ± 92	2.8
EFCA	551 ± 12	1254 ± 14	2.3
BFCA	490 ± 15	871 ± 10	1.8

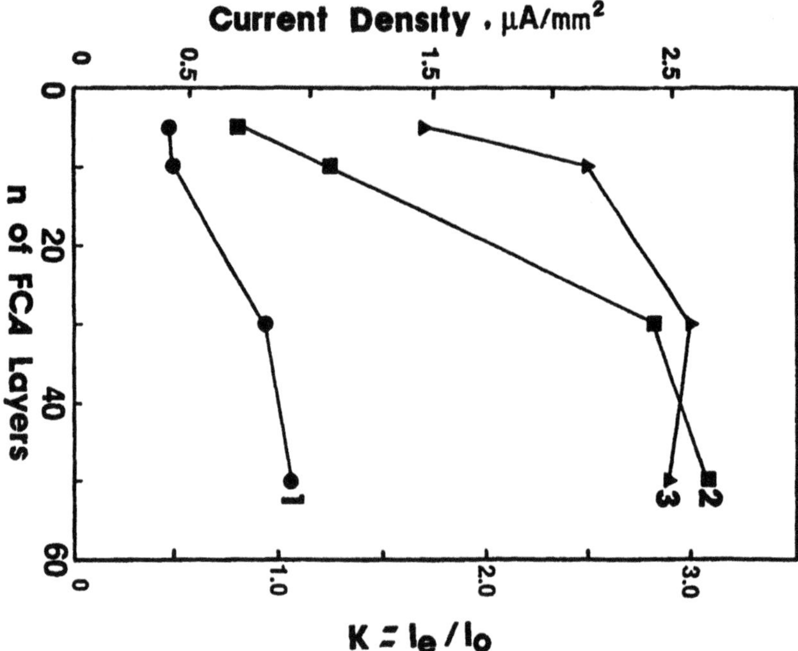

Figure 17. Dependence of the current density on numbers of FCA LB monolayers obtained using abPEI-12 in the absence (**1**) and presence (**2**) 0,1 M glucose. $K = I_e/I_o$ scale for curve 3.

The rate of ferrocene-GOD reaction on the graphite electrode was dependent on the quantity of FCA monolayers. Figure 17 illustrates that the current response to glucose at the modified graphite electrode constantly increases as the number of of FCA layers increases (**2**); the maximum value of $K = I_e/I_o$ was observed for 30 layers. Apparently, a further increase of FCA layers leads to the transition from kinetic controlled to diffusion controlled reactions. The measurement of the ferro-cene-enzyme reaction at 10 layers of FCA was carried out under kinetic conditions.

For the choice of the best conditions for mediator-containing film preparation we formed LB films after mixing of FCA with polymer (3 mg of FCA dissolved in 200 µL abPEI-12 solution in chloroform). Figure 15 (c) displays, the LB isotherm of such a film which has a biphasic character. Analysis of the cyclic voltammograms shows that the addition of GOD and glucose into the solution leads to further increase of the response (from 430 to 1265 nA/mm^2 without and in the presence of GOD and glucose, respectively; $K = 2.9$).

The dependence of the current density upon the glucose concentration was studied for determination of the effective catalytic constants of the enzyme reactions on modified electrode. Figure 18 illustrates the concentration depen-dence; an increase of the current density is observed in the region of kinetic controlled reaction, in accordance with Michaelis-Menten kinetics with the linearity in the range 1-20 mM and with Michaelis constant 16 + 3 mM (a).

Thus, the results obtained indicate that the FD's retain their mediator properties in LB films obtained by the different methods with the help of the positively-

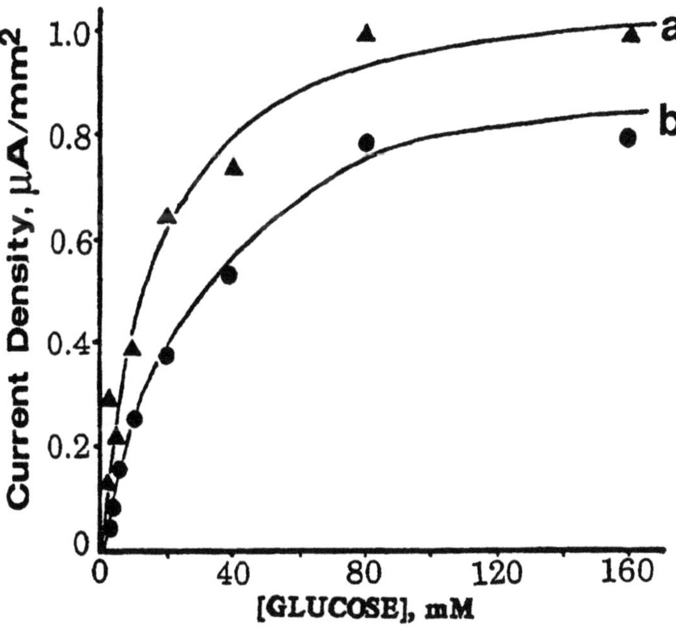

Figure 18. Kinetic curves of the enzyme reaction for dissolved GOD on FCA-abPEI-12 matrix (**a**) and for GOD-FCA-abPEI-12 LB films (**b**). GOD concentration in solution is 10 µM for (**a**); 10 monolayers for (**b**).

charged amphiphilic polymers. Therefore, there was a definite interest in the direct coupling of enzyme to a mediator on an electrode using the described LB technology. So, the multimolecular films obtained on the base of FCA, abPEI-12, and with GOD adsorbed on their surface were transferred onto graphite electrodes. Figure 18 (b) displays the dependence of the amperometric response on glucose concentration. The maximum $K = 2.5$ was observed at 100 mM glucose concentration, and a 1.5-fold increase of the Michaelis constant (27 ± 2 mM) was observed on the immobilization of GOD that was typical for the "immobilized effects." The multilayer glucose sensor did not retain its activity for more than 10 days (upon storage in air and room temperature); the electrode was found to have a useful analytical response for 80 measurements of glucose.

In conclusion, we have demonstrated an approach based on electrostatic interactions of positively charged amphiphilic polymers with the negative charges of ferrocenecarboxylic acid derivatives and glucose oxidase in LB films on the water subphase surface.

Monomolecular films of ferrocenecarboxylic acid, immobilized onto graphite electrodes retain their catalytic properties in the reaction with glucose oxidase. We have shown the possibilities of direct coupling of glucose oxidase and ferrocene-carboxylic acid in multilayer films created by Langmuir technology with the help of amphiphilic polymer matrices.

4.4 Monoamine Oxidase (MAO) LB Films: A New Catalytic Property of the Enzyme

This part of the work was done together with A. Eremenko, A. Barmin, and T. Moskvitina.

LB films of MAO and amphiphilic polyelectrolytes were formed as described above. The MAO activity in LB films calculated as initial rate of the enzyme reaction (V_0) was determined using a Clark electrode in 3.25 ml (in the cell details and construction see Section 4.3). The kinetic parameters of MAO were estimated from the Michaelis-Menten curve and by the method of analysis of the integral kinetic curve of the enzymatic reaction using the ENZFITTER program.

The first experiments on the formation of PVP films included increasing the MAO concentration in the surface layer with and without the polyelectrolyte. For process optimization, several consecutive applications of the enzyme were made and the barrier of Langmuir trough was returned to the start point after each application. After 15 minutes a second compression and following application at a surface pressure of 40 mN/m was carried out.

As shown in Table 7 and Figure 19 the maximal enzymatic activity under the conditions of immobilization without polyelectrolyte appears after the first application (a). After the second application the observed activity decreased to approximately one-third half (b). This correlates with the quantitative measurement of protein in the Langmuir trough before and after the first application: from 0.54 mg

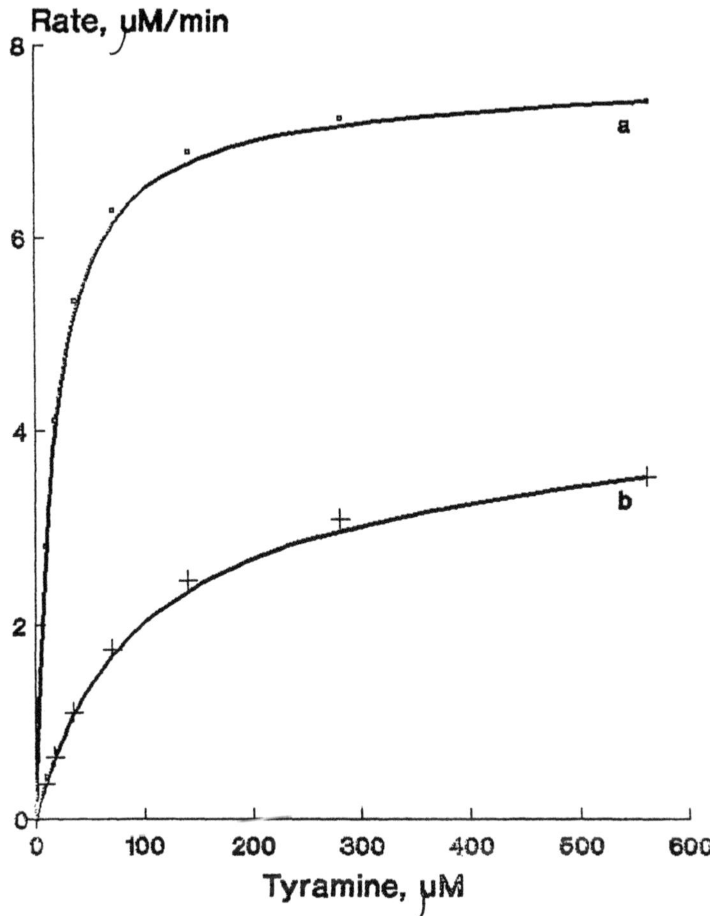

Figure 19. Dependence of the enzymatic reaction rate on substrate concentration on application of MAO without polyelectrolyte. (**a**) First application, (**b**) second application.

of protein after the immobilization, 0.16 mg of it was left on the first three membranes, e.g. less than 30%. In the case of immobilization of the enzyme with abPEI-12, after the first application one could observe the inhibition of MAO by the substrate [Figure 20 (a)]. After formation of a monolayer of the enzyme this effect disappears [Figure 20 (b)]. But Vm in the first case is obviously higher than in the second case of application.

For the quantitative comparison of characteristics of the enzyme immobilized on the electrode membrane the dependence of V_0 initial rate of O_2 decreasing on substrate (tyramine) concentration was studied. The results of the measurements

Table 7. Kinetic Constants of the MAO in the LB Films

Application Type	Tyramine Michaelis Constant (Km), M	Maximal Rate (Vm), μM/min	Vm/Km, min^{-1}
Pure M AO:			
1st application	15.2 ± 2.4	7.6 ± 0.2	0.5
2nd application	95.4 ± 6.6	4.1 ± 0.1	0.04
MAO with abPEI12:			
1st applicaion	36.0 ± 12.0	12.0 ± 1.8	0.3
2nd application	189.3 ± 7.4	7.5 ± 0.1	0.04

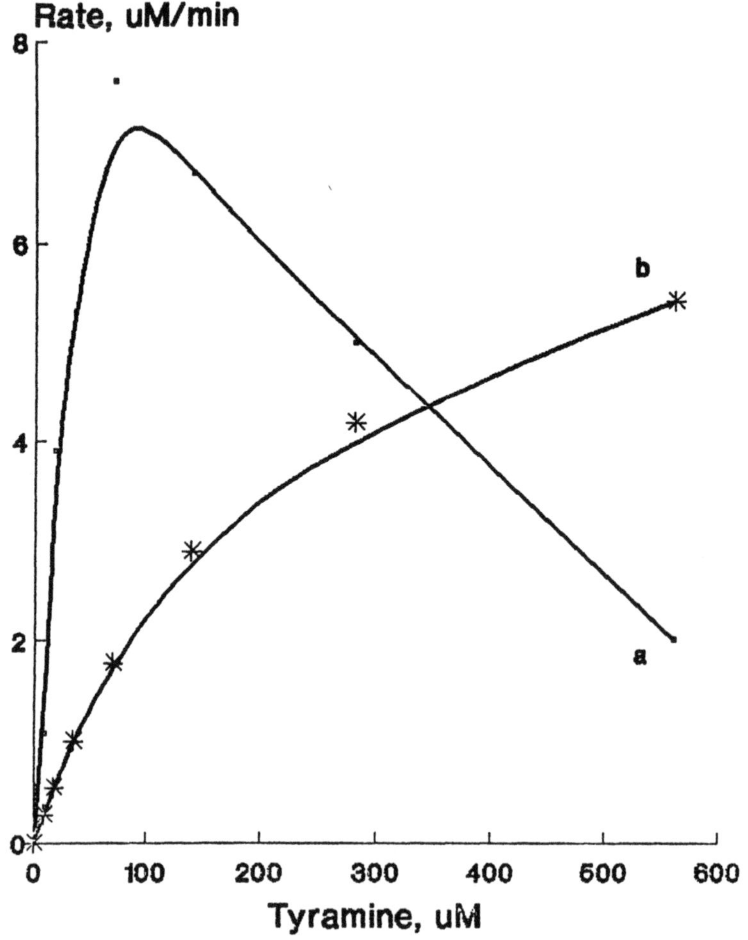

Figure 20. Dependence of the enzymatic reaction rate on substrate concentration on application of MAO with abPEI-12. (**a**) First application, (**b**) Second application.

of these concentration dependences for five modifications of application: with pure MAO and with different polyelectrolytes—are seen in Figure 21 and Table 7.

Analyzing this data one may emphasize three essential important moments:

1. The minimal meaning of the Michaelis constant and the maximal ratio Vm/Km was observed under the immobilization of MAO without polyelectrolytes.
2. After the application of the enzyme in the presence of all electrolytes except abPEI-12 one may observe increasing maximal velocity [Figure 21

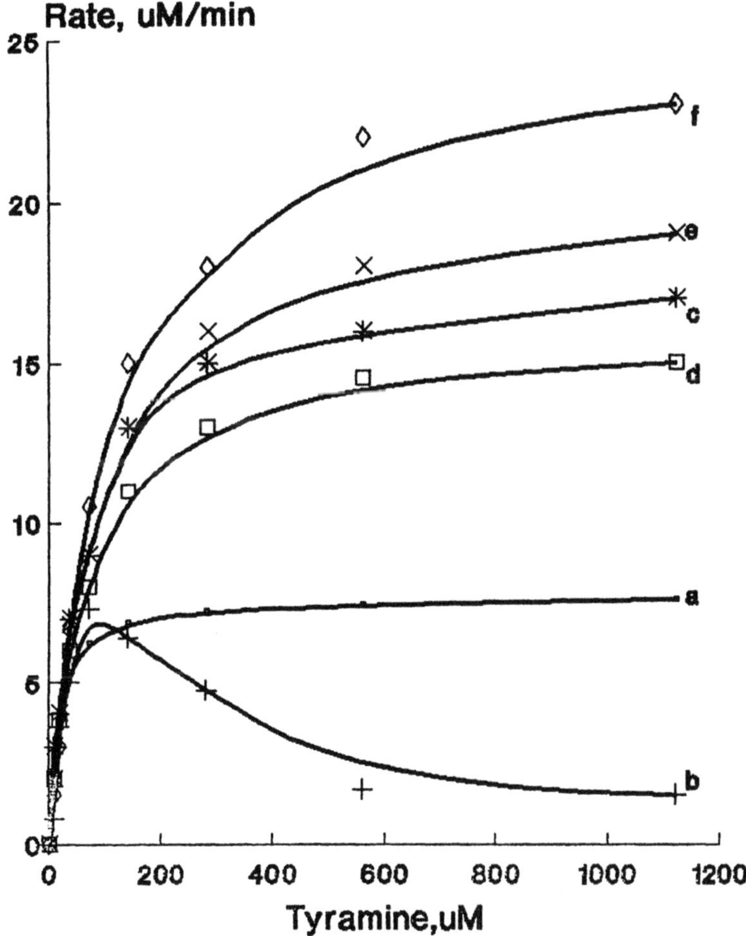

Figure 21. Dependence of the enzymatic reaction rate on substrate concentration with different modes of MAO immobilization. (a) Without polyelectrolyte, (b) with abPEI-12, (c) with abPEI-70, (d) with alPEI- 70, (e) with alPEI-35, (f) with aP4VP-70.

(c-f), Table 8], the most distinct using PVP-70 [Figure 21 (f), Table 8], accompanied by strong increasing apparent Michaelis constant, i.e. . Vm/ Km decreases. This indicated an increase in concentration of the enzyme in the layer in the presence of polyelectrolyte.

3.   Using abPEI-12 as a amphiphilic polyelectrolyte one can observe a sharply declining Michaelis kinetic, most accurately described with the model of allosteric surplus substrate inhibition.

There are data in the literature that serotonine deaminative activity may be inhibited by substrate under, for example, MAO aggregation or as result of definite quantity of lipid presence. In connection with this was supposed the possibility of allosteric regulation of MAO by serotonine, but not by tyramine[24]. In this experiment the possibility of allosteric regulation of the activity of MAO by tyramine was demonstrated.

Using the program of nonlinear regression, ENZFITTER, we analyzed the integral kinetics of MAO, immobilized with abPEI-12 and without any amphiphilic polyelectrolytes (see Table 9).

These data indicate that the effective immobilization of MAO without polymer is possible, however, the presence of amphiphilic polyelectrolytes on the air-water interface led to considerable surface concentration of the enzyme that may be very important in the technology of an enzyme immobilization during biosensor fabrication.

To evaluate the relative stability of the enzyme films their 1-day and 7-day properties after enzyme immobilization at 280 μM substate concentration were

Table 8.   Kinetic Constant of MAO in the LB Films with Amphiphilic Polyelectrolytes

Polymer Type	Tyramine Michalis Maximal		VM/Km, min⁻¹
	Constant (Km), μM	Rate (Vm), μM/min	
pure MAO	15.2 ± 2,4	7.8 ± 0.2	0.5
abPEI-12	36.0 ± 12.0	12.0 ± 1.8	0.3
abPEI-70	55.0 ± 5.8	17.8 ± 0.5	0.3
alPEI-70	60.0 ± 5.3	16.0 ± 0.7	0.2
alPEI-35	77.5 ± 2.1	20.6 ± 2.1	0.2
aP4VP-70	89.3 ± 18.6	24.4 ± 1.4	0.2

Table 9.   Kinetic Constants of MAO in the LB Films Obtained from Integral Kinetic Curve at Saturation of Tyramine Concentration

Type of Film	Oxygen Michael is Constant (Km), μM	Maximal Rate (Vm), μM/min	Vm/Km, min-1
Pure MAO	6.0 ± 0.6	8.5 ± 0.2	1.4
MAO with abPEI-12	15.1 ± 0.1	6.0 ± 0.5	0.4

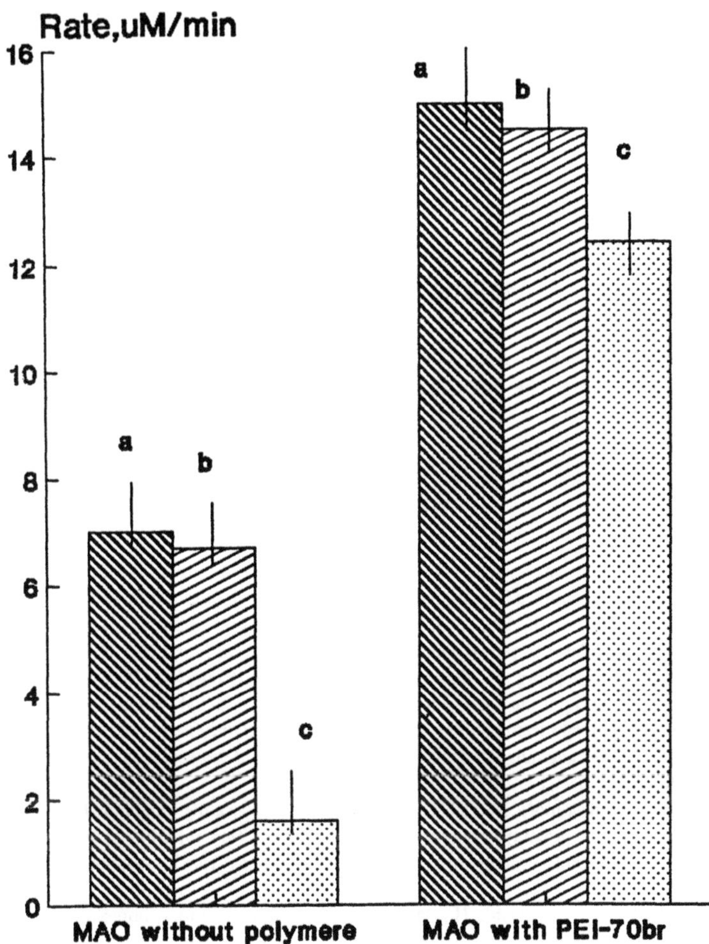

Figure 22. Stability of the enzyme in LB films without (**A**) and with (**B**) abPEI-70 (**a**) Base level, (**b**) after one day, (**c**) after 1 week; tyramine concentration 280 µM.

compared. Figure 22 illustrates MAO activity of these films in the absence (A) and in the presence (B) of abPEI 70. Changes in the enzyme activity were not obtained on the first day for in cases [Figure 22 A(b), B(b)]. However, the activity of immobilized MAO without polyelectrolyte was only 20-25% [Figure 22A(c)] after 1 week storage in air at room temperature, while in the case with polyelectrolyte it was 80-85% [Figure 22B(c)].

In conclusion, we have demonstrated Langmuir-Blodgett films of lipid-dependent MAO on a water surface. Using amphiphilic polyelectrolytes led to the formation of protein-polymer complexes on the air-water interface, and the compresion of such complexes further led to the increase of surface concentration and the change of catalytic properties of the enzyme. Furthermore, it was shown that the use of amphiphilic polyelectrolytes led to increased enzyme stability in monomolecular films immobilized onto an electrode. Thus, the data obtained in this work indicate the universal approach of protein-polymer film formation with the help of amphiphilic polyelectrolytes on an air-water interface and their immobilization onto biosensor surfaces.

ACKNOWLEDGMENTS

The author gratefully acknowledges the contributions made by coworkes whose names are given in the article text. The finest thanks go to Mrs. I. Uvarova for invaluable help and technical assistance.

REFERENCES

[1] Miyauch, S., Arisawa, S., Arise, T., Yamamoto, R. (1989). Study on the concentration of an enzyme immobilized by Langmuir-Blodgett films. *Thin Solid Films*, **180**, 293-298.

[2] Furuno, T., Sasabe, H., Ulmer, K.M. (1989). Binding of ferritin molecules to a charged polypeptide layer of poly-1-benzyl-l-histidine. *Thin Solid Films*, **180**, 23-30.

[3] Ahlers, M., Blankenburg, R., Grainger, D.W., Meller, P., Ringsdorf, H., Salesse, C. (1989). Specific recognition and formation of two-dimensional streptavidin domains in monolayers: applications to molecular devices. *Thin Solid Films*, **180**, 93-99.

[4] Sugi, M. (1985) Langmuir-Blodgett films—a course toward molecular electronics: a review. *J. Mol. Elect.*, **1**, 3-17.

[5] Nagagawa, T., Kakimoto, M. (1991). New method for fabricating Langmuir-Blodgett films of water-soluble proteins with retained enzyme activity. *Thin Solid Films*, **202**, 151-156.

[6] Nagayama, K. (1992). Protein array: an emergent technology from biosystems. *Nanobiology*, **1**, 25-37.

[7] Sriyudthsak, M., Yamagishi, H., Moriirumi, T. (1988). Enzyme—immobilized Langmuir-Blodgett film for a biosensor, *Thin Solid Films*, **160**, 463-469.

[8] Arisawa, S., Arise, T., Yamamoto, R. (1991). Concentration of enzymes adsorbed onto Langmuir films and characteristics of urea sensor. *Thin Solid Films*, **207**, (in press).

[9] Blankenburg, R., Meller, P., Ringsdorf, H., Salesse, C. (1989). Interaction between biotin lipids and streptavidin in monolayers: formation of oriented two-dimensional protein domains induced by surface recognition. *Biochemistry*, **28**, 8214-8221.

[10] Tatsuma, T., Tsuzuki, H., Okawa, Y., Yoshida, S., Watanabe, T. (1991). Bifunctional Langmuir-Blodgett film for enzyme immobilization and amperometric biosensor sensitization. *Thin Solid Films*, **202**, 143-150.

[11] Owaku, K., Shinohara, H., Ikariyama, Y., Aizawa, M. (1989). Preparation and characterization of protein Langmuir-Blodgett films. *Thin Solid Films*, **180**, 61-64.

[12] Margolin, A.L., Izumrudov, V.A., Shvedas, V.K., Zezin, A.B., Berezin, I.V., Kabanov, V.A. (1981). Preparation and properties of penicillin amidase immobilized in polyelectrolyte complexes. *Biochim. Biophys. Acta*, **660**, 359-365.

[13] Margolin, A.L., Sherstyuk, S.F., Izumrudov, V.A., Zezin, A.B., Kabanov, A.V. (1985). Enzymes in polyelectrolyte complexes. The effect of phase transition on thermal stability. *Eur. J. Biochem.*, **146**, 625-632.

[14] Zezin, A.B., Izumrudov,V.A., Kabanov,V.A. (1989). Interpolyelectrolyte complexes as a new family of enzyme carries. *Macromol. Chem., Macromol. Symp.*, **26**, 249-264.

[15] Sukhishvili, S.A., Nikolaychik, D.V., Polynsky, A.S., Yaroslavov, A.A., Chechik,O.S., Kabanov, V.A. (1989) Competitive interactions in poly-*N*-ethyl-4-vinylpyridinium bromide-bovine albumin-carboxylated latex system. *Immunologia*, 82-84 (Russian).

[16] Yaroslavov, A.A., Polynsky, A.S., Sukhishvili, S.A., Kabanov, V.A. (1989). The interaction between artificial antigens based on synthetics polyelectrolytes with immune system cells: physico-chemical aspects. *Macromol. Chem., Macromol. Symp.*, **26**, 265-280.

[17] Mustafaev, M.i., Blokhina, V.D., Agafieva, V.S., Kabanov, V.A. (1980). On competitive relation of protein fractions of serum in formation of their complexes with polycations. *Molekularnaya Biologiya*, **14**, 64-75.

[18] Dzantiev, B.B., Blintsov, A.N., Zivileva, L.S., Berezin, I.V., Egorov, A.M., Izumrudov,V.A., Zezin, A.B., Kabanov, V.A. (1988). Interaction of the antigens with the antibodies immobilized on a synthetic water-soluble polyelectrolytes. *Doklady Akad. Nauk SSSR*, **302**, 222-225.

[19] Kabanov, V.A. (1986). Synthetic membrane active polyelectrolytes in desing of artificial immunogenes and vaccines, *Macromol. Chem., Macromol. Symp.*, **1**, 101-124.

[20] Bergmeyer, H.U., Grassl, M., Walter, H.E. (1989). *Methods of Enzymatic Analysis*, 3d edn., Vol. 2. p. 201. VCH, Weinheim.

[21] Carr, P.W., Bowers, J.D. (1980). *Immobilized Enzymes in Analytical and Clinical Chemistry*. Wiley, New York.

[22] Guilbault, G. (1984). *Analytical and Clinical Chemistry*. Wiley, New York.

[23] Tuner, A.P.F., Karube,I., Wilson,G.S. (Eds.) (1987). *Biosensors. Fundamentals and Applications*. Oxford University Press, New York.

[24] Moskvitina T.A., Kuchina N.E. (1982). Kinetic properties of multiply isoforms of MAO from bovine brain. *Voprosy medizinskoy chimii*, **28**, 127-131.

A FAMILY OF SENSORS BASED ON THE PORPHYRIN PHOSPHORS

Dmitry B. Papkovsky

OUTLINE

Advances in Biosensors
Volume 3, pages 111-126.
Copyright © 1995 by JAI Press Inc.
All rights of reproduction in any form reserved.
ISBN:1-55938-535-9

ABSTRACT

Recent achievements of our group specializing in luminescent sensors are described. The development of new advanced probes (phosphors) and derived solid-state sensing materials gave rise to some new approaches to fiber-optic quenched-luminescent sensing, corresponding instrumentation, arrangements, and special optoelectronic device(s). A number of practical sensing systems which include: fiber-optic lifetime-based oxygen sensor; fiber-optic lifetime-based enzyme biosensor; flow-cell enzyme sensor for phenols; and lifetime-based sensor for relative air humidity and temperature sensor are presented. This all might be considered as an integral block of scientific and technical production of "multi-funtional sensors," or a family of sensors. The course of the research, present status and future prospects are discussed.

1. INTRODUCTION

Extensive studies in the field of sensors and, in particular, in optical (bio)sensors have been performed during the last decade [1, 2]. This is the result of vital practical needs of industry, medicine, and ecological control etc. for such analytical systems and devices. These needs continue to increase at a rapid pace.

Luminescent probes combine high selectivity and sensitivity of detection together with sensitivity to changes of various parameters in the system and/or probe microenvironment, such as concentration of the probe, concentration of quenchers, protolytic and complexation reactions, membrane potential, and hydrophobicity. Luminescent sensors present a large and important group in the family of optical sensors.

Two main approaches to luminescence sensing utilize either luminescence intensity or lifetime measurements. Luminescence lifetime is an intrinsic physical parameter of the particular luminophore, hence it is independent of the dye's concentration, scheme of measurement, and long-term drift and fluctuations of the luminescent signal (photodegradation of the dye, instability of the electronic scheme, light source and photodetector, etc.). This permits simple and stable calibration. Luminescence lifetime measurements are usually more complex and have some special requirements of the probe and optical elements, but they are preferred for practical applications.

Molecular oxygen is one of the main quenchers of luminescence [3]. It is of prime importance and the key mediator (effector or substrate) for a variety of derived analytical systems [1, 4]. The development of an effective oxygen sensor (transducer) also provides a basis for a variety of oxygen-mediated sensors (e.g., enzyme biosensors). Other species, such as nitrogen oxides, and heavy metal ions, are also good quenchers, but oxygen continues to dominate the field.

To date, numerous systems for quenched-luminescence sensing of different analytes and corresponding arrangements have been described [1,2,4]. Although they are easily arranged as laboratory prototype sensors, very few were embodied as real commercial sensors and introduced to the market. This phenomenon is explained mainly by bad compatibility between the luminescent probes used and practical requirements of the sensor design. A commercial device, which should be simple, compact, cheap, and reliable, might be constructed on the basis of convenient optical and electronic elements.

Special requirements of sensor design and operation make the use of classic fluorescent probes, such as polycyclic aromatics, FITC, rhodamines, coumarines, in effective in sensors since they need complex measurement device(s) of dramatical complexity, size, and cost. When it became obvious that the first choice of custom probes for sensors was not optimal and fruitful, active studies on design and application of the new probes and approaches to luminescent sensing were undertaken.

Our studies in this area performed during the last 4 years dealt with the development of the porphyrin phosphors, solid-state sensitive coatings and quenched-luminescence sensor(s). Special emphasis was given to quenching with oxygen, fiber-optic detection, lifetime-based phosphorescence measurements, and the construction of prototype sensors.

2. ADVANCED SENSING MATERIALS

2.1 Porphyrin Phosphors

Some of metalloporphyrin dyes (see Figure 1) which have intense absorbance bands in the near-UV and visible region can exhibit intense red phosphorescence at room temperature [5, 6]. Mainly, short-decay Pt(2+)-and Pd(2+)-porphyrins are the most attractive. Characteristic features of these dyes which differentiate them from the classic probes are: (1) intense longwave excitation band(s) in the spectral range above 500 nm; (2) intense red emission having Stokes' shift of above 100 nm and high quantum yields of 0.1-0.6; and (3) submillisecond time-scale of luminescence decay kinetics.

Such luminescent properties make metalloporphyrin dyes promising for use in optical sensors—in particular, for quenched-luminescence oxygen sensing by lifetime measurements [7-9]. Water-soluble derivatives were shown to be applicable as intrinsic probes for oxygen sensing in solutions [7,9,10].

On the other hand, the solid-state sensing approach is much more advantageous [11] and widely used in sensors. We tested the porphyrin phosphors in the solid-

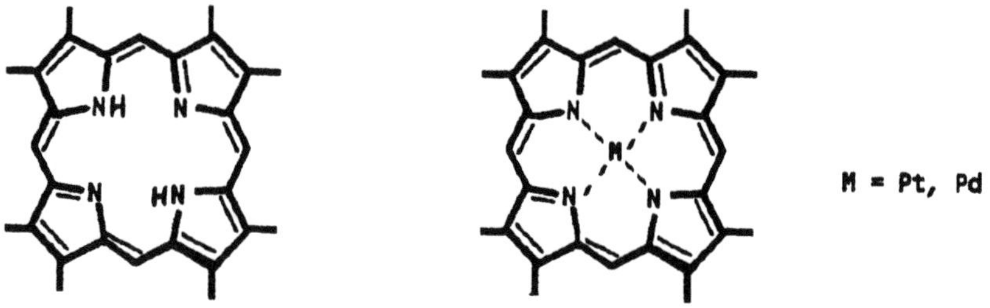

Figure 1.   Structures of the porphine ring (I) and its metallocomplex (II).

state approach. The following basic types of sensitive coatings have been developed:

- phosphorescent oxygen-sensitive polymer compositions (*thick polymer films*) on the basis of hydrophobic porphyrin derivatives, and

- phosphorescent mono- and multilayer sensitive coatings based on Langmuir-Blodgett (LB) assemblies (*thin solid films*) and amphiphylic (water-soluble) dyes.

2.2 Phosphorescent Polymer Compositions

Detailed research was performed to study the behavior of porphyrin phosphors in solid polymer matrices [12]. Several types of polymers were tested to select the optimal one for quenched-phosphorescence oxygen-sensing applications.

As a result, a special paint was developed. It comprises a solution of a hydrophobic phosphor and polystyrene matrix base dissolved in an appropriate solvent. The paint was applied onto solid surfaces and supports to produce after drying the solid-state phosphorescent composition: thick polymer film coating. A suitable brush or spraying could be used for the application.

The paint was applied onto a suitable solid support, such as a polyester transparent film or fiber-optic tip. These sensor-active elements (oxygen membranes) have improved mechanical properties and are easily handled [13].

Thick polymer films thus obtained possessed some marked advantages in comparison with the existing oxygen-sensitive coatings. These concentrated true solid solutions of the dye(s) possess high long-decay luminescent signals, sensitivity to oxygen and quick response. Moreover, a decay curve was described satisfactorily by a single exponential decay law, and quenching with oxygen proceeded according to simple Stern-Volmer dependence [11]. This is very useful for appli-

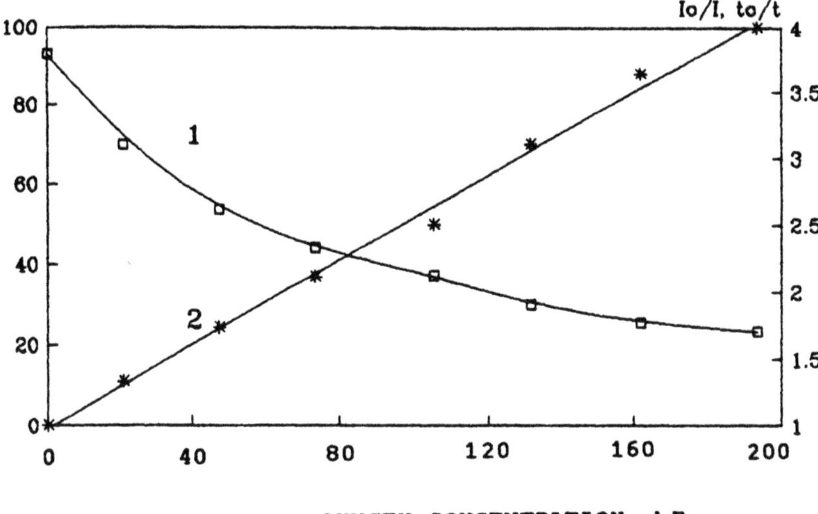

Figure 2. Standard curve for the lifetime-based oxygen sensing with a Pt-octaethylporphine-polystyrene thick film: lifetime vs oxygen plots. (Left) Y-axis; Stern-Volmer plots. (Right) Y-axis.

cation in lifetime-based oxygen sensors. Predetermined calibration which is quite stable with time or single-point recalibration could be used. A typical standard curve for the lifetime-based oxygen sensing is presented in Figure 2.

A 10-fold difference in luminescence lifetime values between Pt-porphyrins and Pd-porphyrins (non-quenched lifetime values of about 0.1 ms and about 1.0 ms respectively) and for their polymer compositions leads to markedly different sensitivity to oxygen. The first type is optimal for sensing of normal oxygen ranges between 1 and 20 kPa or 10 and 250 μM (see Figure 2). The second type is much more suitable for the low-oxygen (anaerobic) control since at ambient oxygen they appear to be almost totally quenched and the luminescent signal is weak. They both have similar physical properties and together give a complementary pair which cover a broad analytical range of oxygen concentrations.

After detailed research on studying and optimization with thick polymer films, they were introduced to practical sensing applications.

2.3 Phosphorescent LB-Coatings

Another technique for preparation of solid-state sensitive coatings was based on the monomolecular assemblies containing porphyrin phosphors. Langmuir-Blodgett technology is promising for sensing applications since it is capable of producing regular solid-state coatings with predetermined and highly reproducible properties [14]. This is especially vital for the disposable sensing elements.

In the previous studies, the phosphorescent LB-coatings were prepared by the classic LB-technique [15]. Hydrophobic porphyrin molecules were incorporated passively into lipid LB-structures. But the coatings thus obtained suffer from aggregation of the dye molecules, decreased luminescent signals, and complex mechanism of quenching with oxygen. Their applicability to practical sensing seems questionable.

We developed a special technique for constructing surface assemblies and preparating phosphorescent LB-coatings [16]. It uses amphiphylic (water-soluble) dyes (such as polycarboxylic metalloporphyrins) which participate actively in the formation of oriented interface structures—thin solid films. In some cases, special amphiphylic-charged polymers (such as alkylated polyethyleneimine) was used to improve formation of LB-coatings with high concentration of the dye. The principle of the latter technique is described in detail elsewhere [16,17].

Mono- and multilayer thin solid film coatings obtained in this manner also gave relatively high and stable phosphorescent signals, but their quenching with oxygen proceed in a strictly different manner. Contrary to the thick polymer films [12], the new structures possessed markedly decreased sensitivity to oxygen. Although phosphorescence decay was simple, single exponential quenching with oxygen was found to be dependent on certain environmental parameters—in particular, water content. In a dry air atmosphere almost no quenching with oxygen was observed, even for the long-decay Pd-porphyrins [16], whereas in aqueous solutions, or in wet air, oxygen quenching is restored. Such a modulation of quenching was shown to be reversible and resulted in both intensity and lifetime changes.

Although specific phosphorescent signals of the thin solid film coatings were high, their absolute values were much lower than for the thick polymer films. For 10-layer coatings the difference was about 10-fold. Weak signals may cause difficulties in fiber-optic sensing, especially by lifetime measurements.

2.4 Phosphor Design

Although the above solid-state sensitive coatings based on Pt- and Pd-porphyrins displayed a number of advantages, serious problems dealing with constructing simple measurement devices remain unsolved. The main problem is to find an effective semiconductor light source for the sensor, such as LEDs or lasers.

In particular, the longwave excitation band of Pt-porphyrins is sharp and has an emission maximum at about 535 nm. They are weakly excitable with convenient narrow-band green and pure-green LEDs (emission maximum at 567 and 557 nm respectively [18]). Blue LEDs which have broad emission spectra but low energy output, are also ineffective. Pd-porphyrins (excitation maximum at 545 nm) are somewhat more excitable with LEDs, but the phosphorescent signals obtained are also weak. This situation is critical for the fiber-optic sensors since high losses of

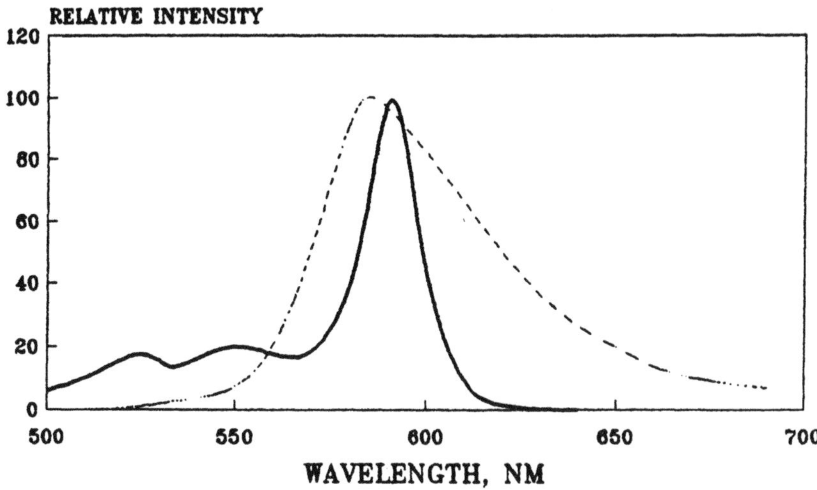

Figure 3. LED-compatibility of the advanced phosphor: (——) absorbance (excitation) spectrum of the dye; (-----) emission of the yellow LED.

both excitation and emission light could not be avoided. As a result, complex and expensive optics and electronics (powerful lamps, interference filters, sensitive photodetectors or photomultiplier tubes and amplifiers) needed to be used to obtain high enough luminescent signals and signal-to-noise ratio. Perhaps the future development of optoelectronic materials could be considered, but we felt this is not a reliable way to solve the particular problem. Our decision was to improve the phosphors used.

A detailed research study was performed based on organic chemistry and spectral methods to create new dyes with more optimal phosphorescent spectral characteristics. The study resulted in a new class of phosphors. In comparison with Pt-porphyrins, the new phosphors have similar phosphorescent efficiency, lifetime, and quenching characteristics. The main difference was in longwave-shifted excitation and emission bands—the shift being about 50 nm and about 130 nm, respectively. High absorbance in the yellow-red spectral region make the new phosphors highly excitable with custom LEDs.

One of the derivatives of the new phosphors similar to Pt-porphyrin was tested in quenched-luminescence sensing experiments. A Fiber-optic lifetime-based

Table 1. The Main Luminescent Characteristics of Porphyrin Phosphors (20 ° C)

Phosphor	*Excitation*	*Emission*	*Lifetime*	
	max., nm	*max., nm*	$\tau_0{}^*$, μs	$\tau_q{}^*$, μs
Pt–porphyrin	535	650	93	21
Advanced Phosphor	592	759	64	15

Source: τ_0, τ_q—Non-quenched and quenched with ambient air phosphorescence lifetime values.

oxygen sensing approach with thick polymer films was used. Spectral compatibility of the new sensitive coating with yellow LEDs is shown in Figure 3. The main phosphorescent characteristics of the two oxygen-sensitive polymer compositions are presented comparatively in Table 1. This new generation of phosphors and corresponding solid-state sensitive coatings are promising for use in fiber-optic lifetime-based oxygen sensors to replace the metalloporphyrin dyes. Only minor modifications of the basic technologies of active elements preparation and assay protocols are needed.

3.   THE DEVICE DESIGN

3.1   Basic Research Instrument

The arrangement for quenched-luminescence sensing experiments was constructed on the basis of an LS-50 luminescence spectrometer (Perkin Elmer, U.K.) equipped with an R928 red-sensitive PMT and supplied with a fiber-optic output [11,12].

The device has a 50 Hz, 8.3 W discharge xenon lamp as a pulsed light source and a special option for the stepwise time-domain measurement of phosphorescence decay curves. The experimental data obtained was then processed with a suitable software ("Enzfitter", Biosoft) to calculate the lifetime value. The minimal time resolution step of the device of 0.01 ms allows precise measurement of lifetime values of about 5 microseconds and above. This is quite optimal for the time-resolved studies of the phosphors and sensitive compositions under investigation, and for the lifetime-based sensing experiments.

The fiber-optic unit comprised a 3-mm bundle made of glass 25-mm monofibres divided at one end into two equal arms and supplied with a light-protecting core and ferrules. Monochromatic excitation beam of an LS-50 was guided through the input arm to the phosphorescent sample (solid-state coating) placed at the common end and its luminescence collected and guided back through the output arm and LS-50 emission optics to PMT. This fibre-optical configuration is, of course, not optimal and high losses of the sample luminescence occur. But since initial signal was usually high, there were no difficulties dealing with lack of sensitivity or small S/N ratio. Solid samples with of 1 to 2 orders weaker luminescense than standard Pt-porphyrin polymer coatings also could be studied with this device. The scheme of the basic research instrument is presented in Figure 4.

The instrument allows one to perform luminescent intensity, spectral and lifetime measurements, kinetic studies with phosphorescent solid samples, and surface structures which may undergo simple manual manipulations. It was used for all routine experiments on development, study, and optimization of active elements, sensing schemes, and assay protocols.

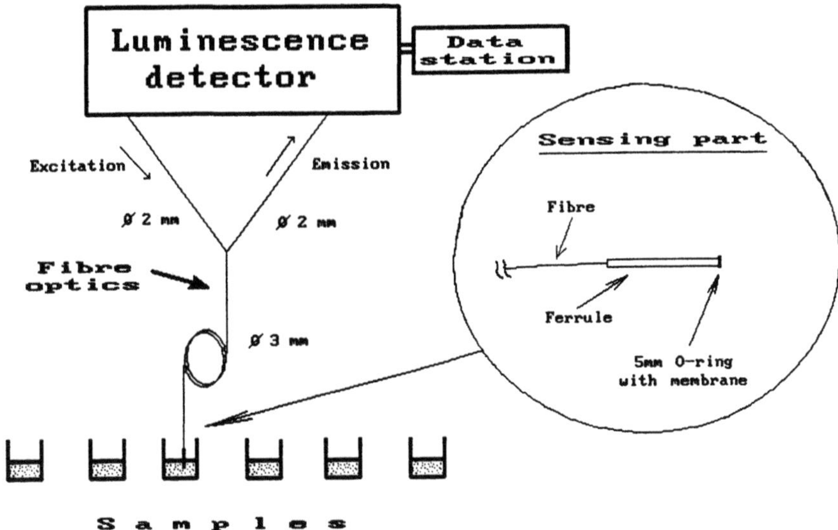

Figure 4. Schematic construction of the basic research instrument.

3.2 Optoelectronic Device

Research in sensors is primarily aimed at the development of sensing element(s) and specialized simple, compact, and cheap measurement device(s). This goal became a reality with the advanced LED-excitable phosphors, and corresponding sensitive coatings have been created. The device offers the basis for simple customized optoelectronics as well as lifetime-based modifications.

The prototype fiber-optoelectronic device was constructed using a powerful yellow LED as a light source, silicone photodiode as photodetector, and a single broad bandpass filtering of emission light. The optical scheme included the same fibre-optical output unit terminated with the sensing element. Fast optoelectronic elements having characteristic rise and fall times of below 500 ns for LED and below 1500 ns for the photodiode (see e.g. [18]) are suitable for luminescent lifetime measurements in the microsecond time scale.

Although time-domain lifetime measurements are more informative and useful for scientific research, they are not ideal for practical sensing applications. In particular, they have strict requirements to the stability and reproducibility of the light source flashes, thus feedback correction is needed. Sequential multi-point measurements and acquisition which are used for the lifetime calculation are rather complex and time-consuming and not so easy to be performed in the real time scale.

The phase-domain (frequency-domain) measurement approach is more simple and effective. It proved to be advantageous for luminescent lifetime-based sensing [19]. The principles of the two basic methods and analytical dependences are described elsewhere [3].

The frequency-domain measurement approach was selected and used in the lifetime-based device design. For the phosphors having simple single-exponential decay characteristics the lifetime is calculated as,

$$tg(\theta) = 2\pi f\tau, \tag{1}$$

where: τ = luminescence lifetime; f = frequency of the light source modulation; θ = measured phase shift between excitation and emission light modulation. The optimal working frequency of the light source modulation for the frequency-domain measurement device is determined according to the following equation [19]:

$$2\pi f\tau \cong 1 \tag{2}$$

For the oxygen-sensitive polymer coatings with lifetimes ranging from about 15 to 64 ms this value is about 5 kHz. In such conditions they should give the phase shift angles of 25.2° in air and 62.6° in vacuum (calculated values).

With the prototype device constructed and oxygen-sensitive polymer coatings sensitive and stable enough, luminescent signals were obtained. It allowed us to perform both luminescence intensity and lifetime measurements by on-the-flight determination of the phase shifts.

4. PRACTICAL SENSING APPLICATIONS

The two basic types of sensing materials which possess rather different properties with respect to quenching of luminescence with molecular oxygen provided a number of practical sensing applications. The following approaches were realized.

4.1 Fiber-Optic Oxygen Sensor

An insertable membrane-type sensor based on thick polymer films was developed for oxygen sensing [13]. Both luminescence intensity and lifetime measurement principles were used. The sensor is applicable for quantitation and continuous monitoring of oxygen (or air pressure) in liquid and gas phase. Several types of sensitive coatings effective for physiological, anaerobic (low vacuum) oxygen measurements are available. They have satisfactory photo chemical stability and quick response, and are also suitable for flow-through oxygen monitoring, long-distance remote control, and two-dimentional imaging with a fiber-optic probe. Oxygen sensing is performed with chemically inert active elements in a passive manner without consumption of the analyte.

The sensor might be used effectively in the mining, metallurical, chemical, and food industries, and in medicine and ecology (air and water monitoring). It is simple and reliable and has stable calibration in a lifetime mode.

Oxygen-sensitive coatings are easily coupled with oxygen-dependent recognition systems to give biosensors. Two basic types of enzymatic sensors have been developed: fiber-optic membrane-type enzyme biosensor and flow-injection enzymatic sensor with a fiber-optic oxygen detector.

4.2 Fiber-Optic Enzyme Biosensor

Protein molecules were immobilized directly onto a polystyrene-based matrix [20]. The enzyme membranes thus obtained showed sensitivity to the substrate in a sample solution. The latter causes *local* changes of oxygen concentration in the membrane due to enzymatic reaction with the dissolved oxygen which proceeds at its surface and is measured by quenching of phosphorescence. Both luminescence intensity and lifetime measurement principles are applicable for the analyte quantitation.

A reagentless membrane-type glucose biosensor comprises the same oxygen sensor with an oxygen membrane coated with glucose oxidase. It showed high and specific sensitivity by luminescent response to glucose content in the range of 0.05-1.2 mM. A typical response curve and calibration are presented in Figure 5. Good working characteristics (analytical range, precision, working stability, etc.) make the biosensor useful for glucose determination and continuous monitoring in biological fluids, such as blood and urine. General medicine,

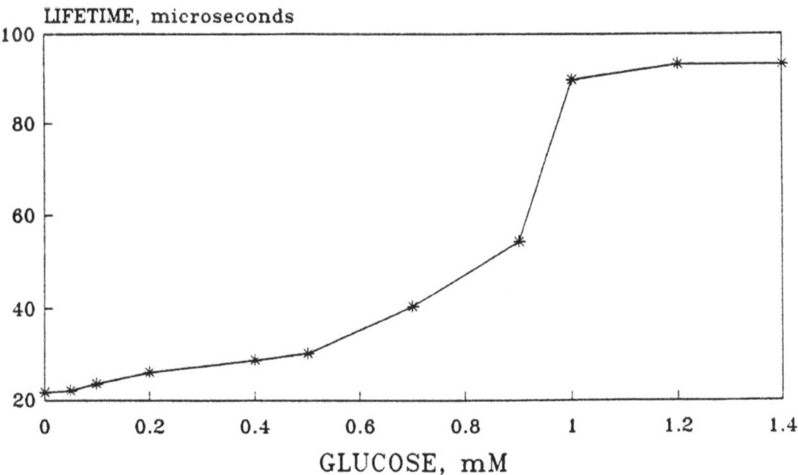

Figure 5. Calibration curve for the lifetime-based fiber-optic glucose biosensor.

home analysis, and biotechnological process control, are potential applications
for this unit.

4.3 Flow-Cell Enzymatic Biosensor

A fiber-optic oxygen sensor capable of flow-through measurements was
coupled to an enzymatic flow analysis system. A special flow-cell arrangement
was constructed and applied to sensing of (poly)phenols. The integral flow-cell

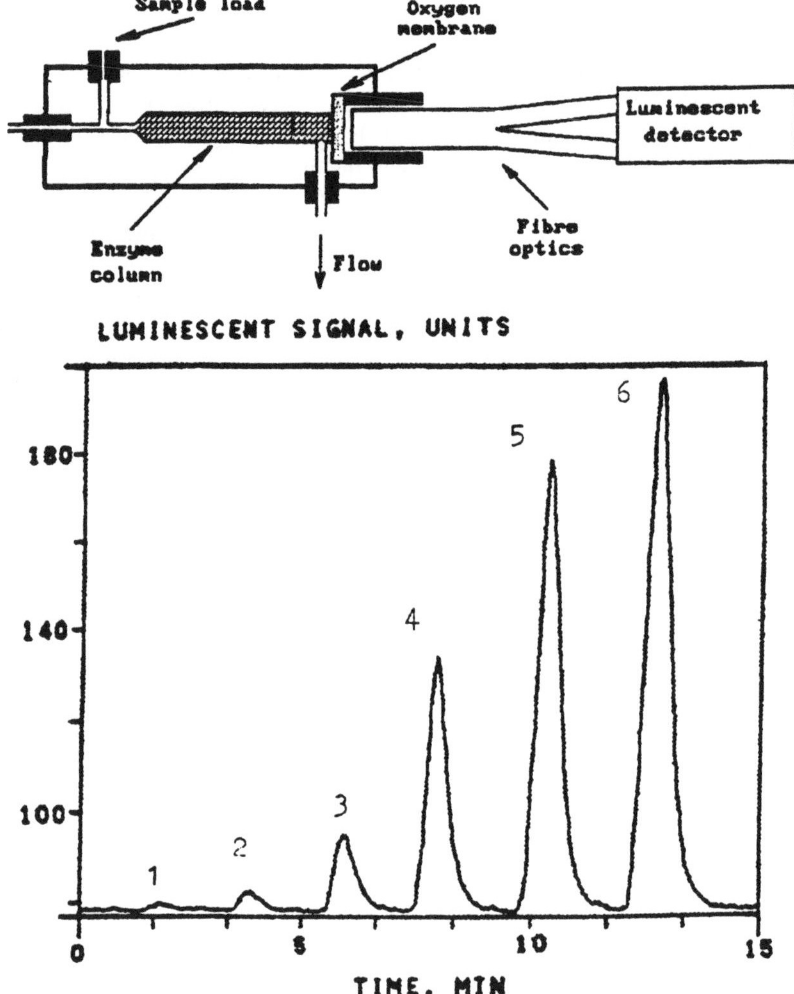

Figure 6. Construction of the enzymatic flow-cell arrangement and typical response
curve to different hydroquinone standards: (1) 0.2 mM; (2) 0.5 mM; (3) 2.0 mM; (4)
5.0 mM; (5) 10.0 mM; (6) 20.0 mM.

unit has a buffer flow path, sample injector, and an immobilized enzyme micro-column, oxygen membrane compartment with a fiber-optic link to the detector. Laccase enzyme was used as an effective recognition system [21]. The sensor was used for the determination of *catechol content* in tea and brandy, and some important *phenols* (phenol, mono- and di-chlorophenols) as well [22]. The sensor scheme and response curve are presented in Figure 6. The system is useful for quality and process control of food products and for industrial wastewater monitoring.

4.4   Sensor for Relative Air Humidity (RH)

A membrane-type gas-phase sensor based on the phosphorescent LB-coatings was developed for the lifetime-based sensing of environmental relative air humidity. Both Pt- and Pd-porphyrin probes are applicable, but Pd-porphyrins are preferred due to their better spectral and lifetime characteristics. Standard curves are presented in Figure 7.

4.5   Temperature Effects and T-Sensing

Quenched-luminescence sensing is temperature-dependent. The main factors which have an influence upon the measured luminescent signal. are: (1) true luminescence lifetime and quantum yield; 2) bimolecular quenching rate constant; (3) solubility of the quencher (oxygen) in the matrix.

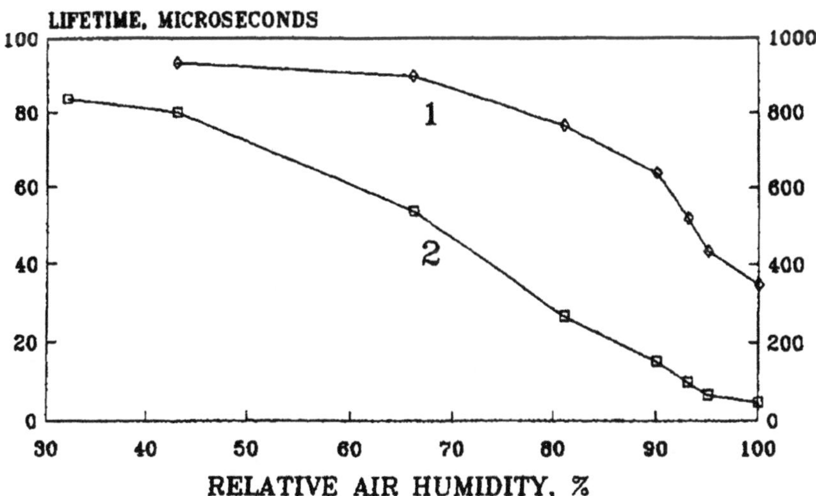

Figure 7.   Standard curve for the lifetime-based RH-sensing with Pt-coproporphyrin III (1) and Pd-coproporphyrin III (2) LB-coatings.

Temperature effects were studied and quantified for the Pt-octaethylporphine thick polymer film coating [12]. The results obtained could be used for T-correction of experimental data (if necessary).

For the LB-coatings the influence of the last two factors might be minor in some cases (low humidity). This made T-dependences simple and dependent only upon intramolecular processes. Although the amplitude of signal changes is not high, the coatings are applicable for the lifetime-based sensing of temperature (in gas phases).

5.   CONCLUSIONS

The scheme presented in Figure 8 shows the course and main steps of the system research performed with the view of developing of practical sensors. It includes the development of the new, highly efficient phosphors based on two basic types of sensitive coatings, a number of important practical applications and a special optoelectronic sensing device(s). In combination, this forms an integral block of advanced materials, technologies, and techniques which might be called a "multifunctional sensor," or a family of sensors. The sensors comprise a basic lifetime-based luminescent fibre-optic detector equipped with a number of replaceable active elements that provide sensitivity to the important parameters and species. The priority character of the main practical results of this study is protected by a number of pending patents.

Most of the sensing applications presented were first realized using Pt- and Pd-porphyrin phosphors and basic research instruments. One could see that these

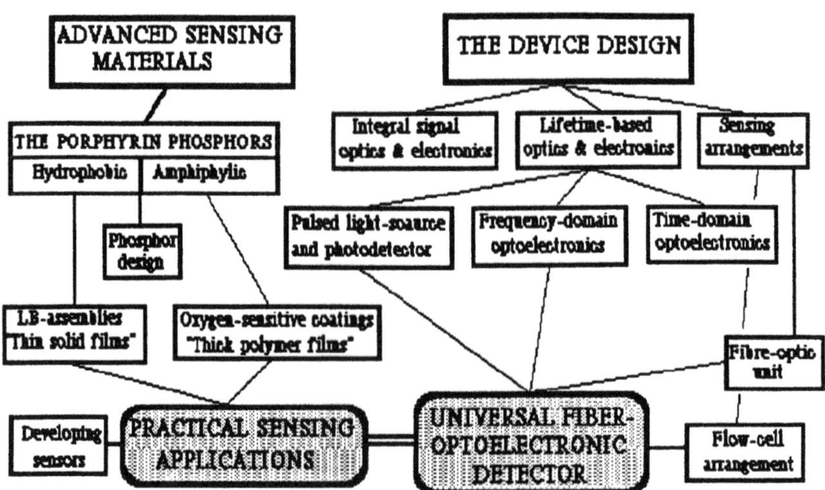

Figure 8.   General scheme of the research in sensors.

phosphors are suitable only for minor practical tasks, but not for the most important, such as fiber-optic sensors for physiological oxygen and enzyme research. At the same time, the systems developed could be easily extended and transferred to the new advanced phosphors and thus coupled with newly created optoelectronic sensing devices.

This work is now in progress. Of course, considerable work is needed to improve and optimize sensing element, the device electronics, and signal acquisition. However, our prototype device, which is capable of quantitative lifetime-based monitoring of luminescence, displays high practical usefulness in solving a variety of analytical tasks. It is believed to compete successfully with the alternative systems now introduced to the market.

ACKNOWLEDGEMENT

The author gratefully acknowledges the contributions made by his coworkers whose names given in the references. The finest thanks are to G.V. Ponomarev from the Institute of Biophysics (Moscow), A.N. Ovchinnikov from the Institute of Eye Diseases (Moscow) and V.I.Ogurtsov from the Power Engineering Institute (Moscow) for their invaluable help and technical assistance.

REFERENCES

[1]   Wolfbeis, O.S. (e.d.) (1991). *Fiber Optic Chemical Sensors and Biosensors,* CRC Press, Boca Raton, Florida.

[2]   Blum, L.J., Coulet, P.R. (1991). *Biosensors Principles and Applications.* M. Dekker, New York.

[3]   Lakowicz, J.R. (1983). *Principles of Fluorescence Spectroscopy, p. 496.* Plenum Press, New York.

[4]   Trettnak, W. (1992). Optical sensors based on fluorescence quenching. In: *Fluorescence Spectroscopy* Wolfbeis, O.S. Ed.) pp. 79-89. Springer-Verlag, Berlin.

[5]   Dolphin, D. (1979). *The Porphyrins.* Vol. 3. Academic Press, New-York.

[6]   Papkovsky, D.B. (1991). Luminescent porphyrin probes. *Appl. Fluoresc. Technol.,* **3**, 16-25.

[7]   Vanderkooi, J.M., Maniara, G., Green, T.J., Wilson, D.F. (1987). An optical method for measurement of dioxygen concentration based upon quenching of phosphorescence. *J.Biol.Chem.,* **262**, 5476-5482.

[8]   Khalil, G.E., Gouterman, M.P., Green, E. (1989). Method for measurement oxygen concentration, *U.S. Patent* No. 4.810.655.

[9]   Papkovskii, D.B., Savitskii, A.P., Yaropolov, A.I. (1990). Oxygen and glucose optical biosensors based on phosphorescence quenching. Anal. Chem. (USSR), **45**, 1441-45.

[10] Papkovskii, D.B., Savitskii, A.P. Yaropolov, A.I. et al. (1991). Flow-injection glucose determination using long-wavelength luminescent oxygen probes. *Biomed. Sci. (London),* **2**, 63-67.

[11] Papkovsky, D.B. (1993). Luminescent porphyrins as probes for optical (bio)sensors. *Sens. Actuat. B.,* **11**, 293-300.

[12] Papkovsky, D.B., Olah, J., Troyanovsky, I.V., Sadovsky N.A. et al. (1992). Phosphorescent polymer films for optical oxygen sensors. *Biosens. Bioelectr.,* **7**, 199-206.

[13] Papkovskii, D.B., Olah, J., Troyanovskii, I.V. Sadovskii N.A. et al. (1991). Fibre-optic oxygen sensor based on phosphorescence quenching. *Biomed. Sci. (London)*, **2**, 536-539.

[14]   Roberts, G. G. (1983). Transducer and other applications of Langmuir-Blodgett films. *Sens. Actuat.*, **4**, 131-145.

[15] Beswick, R.B., Pitt, C.W. (1988). Oxygen detection using Langmuir-Blodgett films of a metalloporphyrin, *Chem.Phys.Lett.*, **143**, 589-599.

[16] Papkovsky, D.B., Ponomarev, G.V., Chernov, S.F., Kurochkin, I.N. (1993). Unexpected luminescent and quenching properties of metalloporphyrins in Langmuir-Blodgett structures. Application to relative air humidity sensing. *SPIE Proc.*, **1885**, 251-257.

[17] Kurochkin, I.N. (this issue).

[18] *Siemens Optoelectronics Data Book 1990.* (1990). Siemens AG, Bereich Halbleiter.

[19] Berndt, K.W., Lakowicz, J.R. (1992). Electroluminescent lamp-based phase fluorometer and oxygen sensor. *Anal. Biochem.*, **201**, 319-325.

[20] Papkovsky, D.B., Olah, J., Kurochkin, I.N. (1993). Fibre-optic lifetime-based enzyme biosensor. *Sens. Actuat. B.*, **11**, 525-530.

[21] Ghindilis, A.L., Gavrilova, V.P., Yaropolov, A.I. (1992). Laccase-based biosensor for detrermination of polyphenols: determination of catechols in tea. *Biosens. Bioelectr.*, **7**, 127-131.

[22] Papkovsky, D.B., Ghindilis, A.L., Kurochkin, I.N. (1993). Flow-cell fibre-optic enzyme sensor for phenols. *Anal.Lett.*, **26**, 1506-1518.

IMMUNOENZYME SENSORS BASED ON AN IMMUNOSORBENT FLOW-INJECTION TECHNIQUE WITH ENHANCED CHEMILUMINESCENCE DETECTION OF A PEROXIDASE LABEL IN A KINETIC REGIME

A. P. Osipov, B. B. Kim, and A. M. Egorov

OUTLINE

Advances in Biosensors
Volume 3, pages 127-142.
Copyright © 1995 by JAI Press Inc.
All rights of reproduction in any form reserved.
ISBN:1-55938-535-9

ABSTRACT

The general principles of flow-injection enzyme immunoassay are discussed. This new approach provides the possibility to perform all stages of the analysis completely automatic and considerably shortens the assay time. To preserve the sensitivity of the assay it is important to have a very sensitive detection system for the enzyme reporter molecule. An enhanced chemiluminescence system was used to detect the peroxidase molecule as a label in flow-injection enzyme immunoassay. A non-equilibrium flow-injection enzyme immunoassay for some model antigens (IgG, tyroxine, 2,4-dichlorophenoxyacetic acid) involving immunoaffinity separation is described.

1. INTRODUCTION

Immunoassays can be defined as quantitative immunological procedures in which the extent of an antigen-antibody reaction is evaluated by label measurement. Such procedures most frequently are based on the use of radioisotopes, fluorescent or enzyme labels of antigens or antibodies, and on the fact that the quantity of a label can be accurately determined by an appropriate technique. Different classes of substances such as hormones, vitamins, proteins, and viral and tumor antigens can be measured using label immunoassays [1].

One of the most widespread approaches in enzyme immunoassay methods is the so called "solid phase" or heterogeneous immunoassay [2]. Solid-phase enzyme immunoassay (EIA) however, is usually a multistep and time consuming procedure. In the first stage of analysis, analyte interacts with antibodies adsorbed on the surface of the solid phase. After washing, one adds a conjugate of the specific antibodies with enzyme label (usually peroxidase or alkaline phosphatase). After the second washing, one measures the enzyme activity on the solid phase which is proportional to the concentration of analyte in a solution.

The most widely used variants of solid-phase EIA are based on the performance of antigen determination in the equilibrium regime. In the case of multistage schemes, each of the stages is carried out for that period of time neccessary to establish equilibrium between the components of the reaction system in solution and on the support. The equilibrium regime of each of the stages enables the greatest participation of the antigen-antibody in the reaction and, consequently, the greatest sensitivity of the analysis to be achieved.

Equilibrium of the reaction between immobilized antibody and antigen in solution is usually reached after several tens of minutes, so the overall time of analysis exceeds 1.5-2 h. Kinetic enzyme immunoassays, in which the products of immunochemical and enzyme reactions are quantitated as initial rates allow a researcher to considerably reduce the analysis time [3,4].

Nevertheless, for the necessary accuracy and reproducibility of the method, the reaction time for each step should be strictly controlled because even small deviations in incubation time at each step can give essentially erroneous results. To this end, flow systems, in which the time for contact of reagents with immunochemical components immobilized on solid supports is well controlled and constant, are very convenient. In recent years, the technique of flow-injection analysis (FIA) has been applied increasingly [5]. This method is based on automatic injection of precisely measured sample zones into a reagent-containing carrier stream. The residence time of the sample in the system is constant. The FIA technique is highly promising for immunochemical assays [flow-injection immunoassay (FIIA)]. However, only a small portion of the rapidly expanding literature on flow-injection methods describes the application of FIA for immunochemical assays. Application of FIA in combination with the kinetic approach in enzyme immunoassay greatly increases the accuracy and reproducibility of analysis.

To achieve high sensitivity in a kinetic immunoassay, the most sensitive detection system for determination of enzyme activity should be used. Enzyme-catalyzed chemiluminescent reactions essentially increase the sensitivity and shorten the quantitation time of enzyme immunoassay. Horseradish peroxidase (EC 1.11.1.7) is widely used as a label in enzyme immunoassay and can be measured with several chemiluminescent procedures. The major advantages of a luciferin or *p*-iodophenol enhanced luminescent procedures are the light levels which are relatively high at pH 8-9 when the background light emission is very low. The assay is rather rapid and sensitive and uses readily available reagents [6,7].

In this chapter we describe some our recent approaches for construction of flow-kinetic immunosensors for quantitation of some model antigens (IgG, tyroxine and 2,4-dichlorophenoxyacetic acid). The inclusion of immunoaffinity columns or solid phases (such as standard polystyrene beads) and enhanced chemiluminescence reactions for the detection of the peroxidase label are the main features of these flow-injection immunosorbent immunoassay systems.

2. EXPERIMENTAL

Reagents. Chemicals, substrates for horseradish peroxidase (HRP), luminol (5′-amino-2,3-dihydro-1,4-phtalazinedione), *p*-iodophenol, thyroxine (T4), Protein A were supplied by Sigma, (U.S.A). 2,4-Dichlorophenoxyacetic acid (2,4-D) were purchased from Riedel de Haen (Germany). HRP, RZ 3.1, was obtained from Agrotecnika, (Ukraine). Acids and salts of analytical grade were Reakhim (Russia) products. Luminol and *p*-iodophenol were recrystallized, the other reagents were used without further purification. All solutions were prepared in Milli-Q water.

Commercial human IgG and rabbit antiserum to human IgG were obtained from N.F. Hamaley Institute of Epidemiology and Microbiology. Monoclonal antibodies against 2,4-D were obtained from Dr. M. Franek (Slovakia). Conjugates of 2,4-D with HRP were prepared using N-hydroxysuccinimide ester [8], while rabbit IgG-peroxidase conjugate was obtained by the periodate method [9].

Immobilization of Antibodies on Polystyrene Beads. The polystyrene beads were incubated in 0.01 mg/ml solution of IgG fraction of rabbit antibodies to human IgG in 0.01 M carbonate buffer, pH 9.6, over night at 4 °C and washed with 0.01 M phosphate buffer (pH 7.4; 0.14 M NaCl, 0.05% Tween 20). Beads were stored at 4 °C, the antibody activity being stable for several months.

Preparation of Substrate Solutions. Luminol and p-iodophenol solutions were prepared by dissolving the exit quantity of reagent in a minimal volume of 0.2 M NaOH and subsequent dilution with 0.05 M Tris-HCl buffer (pH 8.5). H_2O_2 solutions were prepared by dilution of 30% solution with distilled water. The concentrations of luminol and H_2O_2 were verified spectrophotometrically.

Assay of Peroxidase Activity. The chemiluminescent assay of peroxidase activity was performed in a substrate mixture containing 5×10^{-5} M of luminol and p-iodophenol, and 1 mM of H_2O_2 in Tris-HCl buffer, pH 8.6. The intensity of chemiluminescence was measured in relative units. The maximum light intensity was used as a measure of peroxidase activity.

Displacement FIIA Scheme (1) for IgG Determination. The sample was injected into a stream of phosphate buffer (Figure 1), fed through a polyethylene connection tube 1 using a peristaltic pump 2, with the aid of an automatic injector 3; 20-50 µl of the sample was injected. When the sample reached the column with complex antigen-peroxidase-labeled rabbit antibodies against human IgG immobilized on BrCN-Sepharose, pump 2 was stopped for 1-10 min; then pump 2 with buffer and pump 5 with substrate solution (luminol, p-iodophenol, and hydrogen peroxide) were switched on. The immunoenzyme complex displaced from the column mixed with the substrates and one measured the intensity of the chemiluminescence peak of the reaction zone with the help of a flow-cell luminometer, LKB-1251 .

Sandwich FIIA Scheme (2) for IgG Determination. Figure 2 illustrates the flow diagram for the FIIA system using beads with immobilized rabbit antibodies to human IgG and chemiluminescent detection.
A Tecator FIA 5020 instrument equipped with two pumps and a double-injection valve was used. Into the streams of 0.01 M phosphate buffer (pH 7.4; 0.14 M NaCl; 0.05% Tween 20), fed through polyethylene connection tube (1 and 1´) with a peristaltic pump (2) at 0.5 ml/min, with the aid of a double-valve

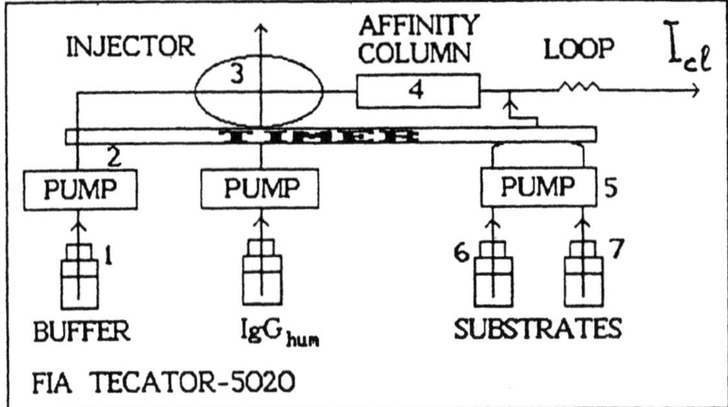

Figure 1.   FIIA manifold for displacement method of enzyme immunoassay for human IgG.

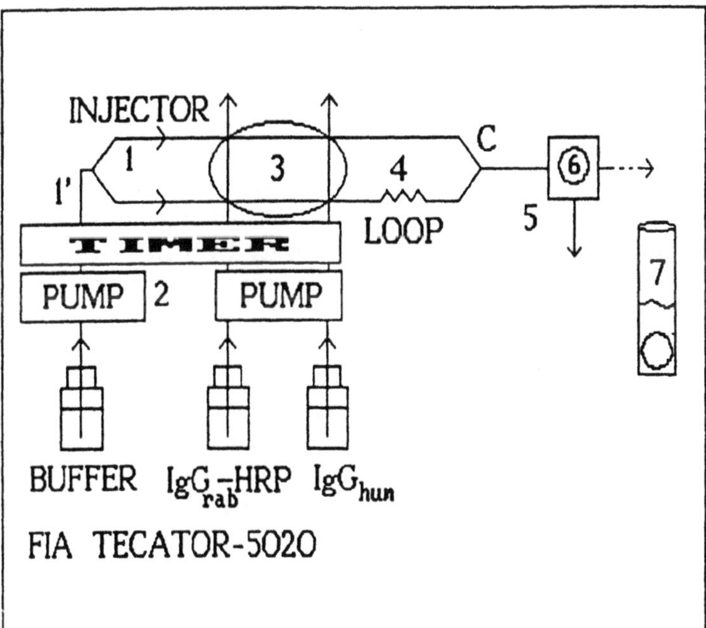

Figure 2.   FIIA manifold for the chemiluminescent sandwich-method of enzyme immunoassay for human IgG.

injector (3), 200 µl of the sample and 200 µl of rabbit IgG-peroxidase conjugate (10^{-7} M) was injected at the same time. The two streams (1 and 1´) were mixed at point C. Through an additional valve (4) (4 m long) point C was reached by the solutions of antigen and antibody-peroxidase conjugate in 1 min. Into a reaction cell (5) a polystyrene bead (6) ("Abbot," 6 mm diameter) with immobilized rabbit antibodies to human IgG was placed. When the antigen sample reached the bead pump 1 was stopped (the time was changed from 1 to 10 min). After switching on pump 1, the bead was washed by the buffer contained in valve 4. When the conjugate sample reached the bead, pump 1 was stopped for the second time (the time varied from 0.5 to 10 min). After the incubation, the immunosorbent was washed by the buffer to remove the unreacted conjugate. Then the bead was transferred from reaction cell 5 to the cell (7) of luminometer LKB-1251, where the chemiluminescence intensity was measured.

FIA Scheme (3) for T4 Determination.   A Tecator FIA 5020 instrument equipped with two pumps and a double-injection valve was used (Figure 3). While assaying T4, both 30 µl of the T4 solution to be analyzed and peroxidase-IgG conjugate were introduced by a two-channel injector (3) in two streams (1 and 1´) of Tris-HCl buffer (0.05 M, pH 8.5, 0.05% Tween-20). A precise adjustment of connecting tubes provided simultaneous mixing of the two streams at point A. The presence of additional loop (4) was dictated by a need for continuous monitoring. When the reaction zone containing the sample with the conjugate was in the loop, pump (2) was stopped for 5, 3, 1.5, or 0.5 min. After switching on pump (2), line reaction zone involving a mixture of the bound and free labeled antibodies was moved onto column (5) with the immobilized BSA-T4 and the pump was again switched off. After incubation for 1 min, the stream was combined (at point B) with solutions of luminol, *p*-iodophenol (7), and hydrogen peroxide (7´) and was driven into a flow cells (8) of a luminometer LKB-1251. The pumps were switched off and the intensity of chemiluminescence was measured for 3 min.

Competitive Enzyme Immunoassay Scheme (4) for 2,4-D Determination. ABICAP (Antibody Immuno-column for Analytical Processes, Abion, Germany) was used as a main unit for immunochromatography (IC). The transparent column contained covalently bound Protein G. The scheme of the assay used was based on sequential competitive enzyme immunoassay procedure [2]. In the first step, free hapten bound to a limited number of antibodies. The unbound antibodies were quantified with a conjugate hapten-HRP in the second step. The routine protocol of the assay was as follows: apply 50 µl of diluted ascitic fluid; wait for 1.5 min; wash time with 500 µl of PBST; add 200 µl of standard solution; wait for 1.5 min; add 50 µl of conjugate solution; wait for 1.5 min; and wash twice with 500 µl of PBST. All steps may be done either manually or automatically.

The activity of the enzyme label was measured by enhanced chemiluminescence. Half a milliliter of substrate mixture was applied onto the column. The column was then placed into a measuring chamber of a luminometer (LKB-1250) and maximal light intensity was recorded 1 min later.

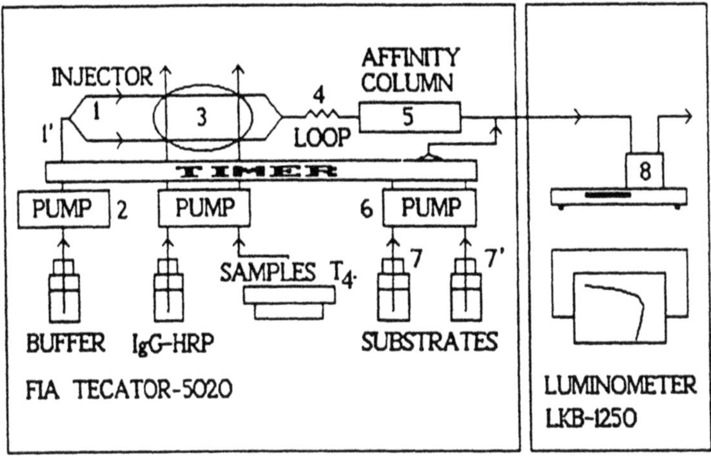

Figure 3. FIIA manifold for the enhanced chemiluminescence immunoassay of thyroxine.

ELISA Experiments. These were performed as follows. Microplates (Nunc-Immunoplates, MaxiSorp.) were coated at 4 °C overnight with Protein A solution, (300 µl, 10 µg/ml). After washing with PBST (3 times) 300 µl of antiserum to pesticide was added in each well of a microplate. After incubation for 2 h at room temperature and washing with PBST, standard (200 µl) and conjugate (50 µl) were added and incubated for 1 h. The wells were washed out and 200 µl of TMB substrate solution (see above) were added. After 15 min of incubation the reaction was stopped with 50 µl of 2 M sulfuric acid. Optical density was measured at 450 nm (reference 650 nm) using a Vmax-reader (Molecular Devices, U.S.). All standards were done in quadruplicate. 2,4-D standards were prepared from 1 mg/ml stock solution in dimethylsulfphoxide. The software package Softmax (Molecular Devices, USA) was used for fitting of standard curves into a four-parameter equation:

$B/B_0 = (1-B_{ex}/B_0)/(1+(C/C_{50})^x) + B_{ex}/B_0$, where C = concentration of analyte; B = concentration of bound tracer; B_0 = concentration of bound tracer at $C \rightarrow 0$; B_{ex} = concentration of bound tracer at $C \rightarrow \infty$; C_{50} = concentration of analyte at which 50% of tracer is bound to antibodies; and x = an empirical coefficient characterizing the slope of the curve at $C = C_{50}$.

3. RESULTS AND DISCUSSION

The concept of flow-injection analysis (FIA) was introduced in its first, simple form in 1974 by J.Ruzicka and E.Hansen [5]. It is based on the injection of a fixed volume of the the sample into a reagent or carrier stream. This results in a sample plug bracketed by reagent. The reagent (or carrier) stream transports the sample to a flow-through detector.

The pump provides a constant flow. This leads to a perfectly constant residence time of the sample in the system. During its travel up to the detector the sample is subjected to mixing with the reagent. The degree of dispersion (or dilution) of the sample can be controlled by varying, for example, the sample volume, the length and diameter of the mixing coils, and the flow rates.

When the dispersed sample zone reaches the detector neither the chemical reaction nor the dispersion process has necessarily come to a steady state. But the experimental conditions are kept equal both for samples and standards; namely constant residence time, constant temperature, and constant dispersion. The sample concentration can thus be evaluated against appropriate standards injected in the same manner as the samples.

We investigated four different possible flow-injection schemes for immuno-assay of some model antigens.

3.1   Displacement FIA Scheme (1)

The first flow-injection scheme for IgG determination we investigated is shown in Figure 1. The principle of analysis is based on the binding of analyte, human IgG, with peroxidase—labeled antibodies partially dissociated from the immobi-lized solid phase complex, and the subsequent registration of the immunocomplex in the water phase by measuring the enzyme activity [4] (Scheme 1).

$$\text{I-IgG*Ab-E}_{\text{solid phase}} \rightleftharpoons \text{I-IgG}_{\text{solid phase}} + \text{Ab-E}_{\text{water}}$$
$$\text{Ab-E}_{\text{water}} + \text{IgG(analyte)} \rightleftharpoons \text{IgG*Ab-E}_{\text{water}}$$
Scheme 1

The calibration graphs for determination of IgG by this method following different times of contact of the immobilized complex of peroxidase-labeled antibody with analyzed antigen are given in the Figure 4. The lower detection limit is about 0.1 mg/ml with a time of incubation of 10 min. A computer simulation of this scheme showed that its theoretical sensitivity is not high and it is necessary to observe the actual relationships between the concentrations of components on solid phase and in solution, and to use low-affinity antibodies.

3.2   Solid-Phase FIA Scheme (2)

The second scheme is a usual two-site enzyme immunoassay of human IgG as a model antigen [10]. Figure 2 illustrates the flow diagram for the FIIA system using standard polystyrene beads (Abbott) with immobilized rabbit antibodies to human IgG and chemiluminescent detection.

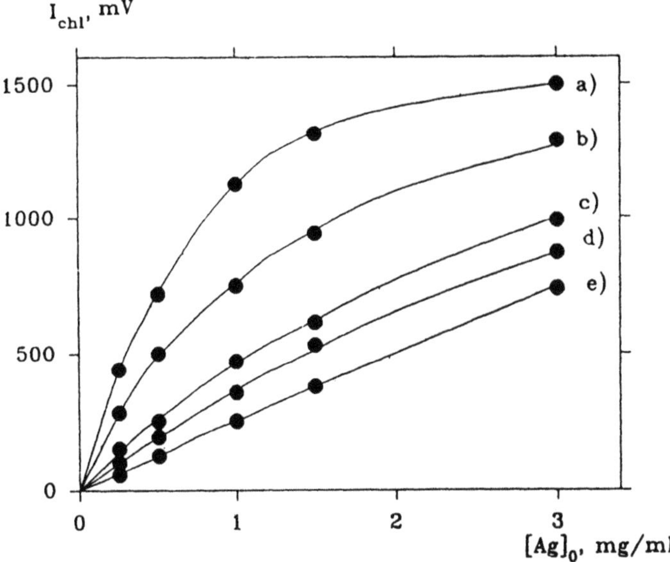

Figure 4. Standard curves of human IgG in flow-injection displacement method of enzyme immunoassay. The times of contact of immobilized complex of peroxidase labeled antibodies with analyzed antigen varied as follows: (**a**) 10 min; (**b**) 5 min; (**c**) 2 min; (**d**) 1 min; (**e**) 0,5 min.

The analytical principle is the same as in the sandwich-immunoassay, the detection limit of antigen is determined by the value of nonspecific adsorbtion of antibody-enzyme conjugate or by the detection limit of the enzyme label. In the kinetic solid-phase enzyme immunoassay, the quantity of immunochemical products on the solid phase is considerably smaller that in the steady-state assay. So it is necessary to use the most sensitive system to detect the enzyme label specifically absorbed on the solid phase. Thus the enhanced chemiluminescence system described earlier in the literature—using luminol, *p*-iodophenol, and hydrogen—peroxide is rather convenient for this purpose. A Tecator FIA 5020 instrument equipped with two pumps and a double-injection valve was used.

After optimization of the conditions for the flow-injection assay, the standard curves for IgG shown in Figure 5 were obtained. The standard curves showed that IgG concentrations of 10^{-9} and 10^{-7} M could be determined at incubation times of 10 and 1 min, respectively (curves a and d). Thus the overall time of assay including two incubation steps and washing can be reduced to about 4 to 5 min.

In summary, the kinetic study of antigen-antibody interactions using flow-injection analysis has allowed us to propose a rapid method of chemiluminescent immunoassay as exemplified by human IgG. Using the advantages of enhanced chemiluminescence we succeeded in carrying out both steps of the sandwich-scheme of the immunoassay in a kinetic fashion, by decreasing the time taken for

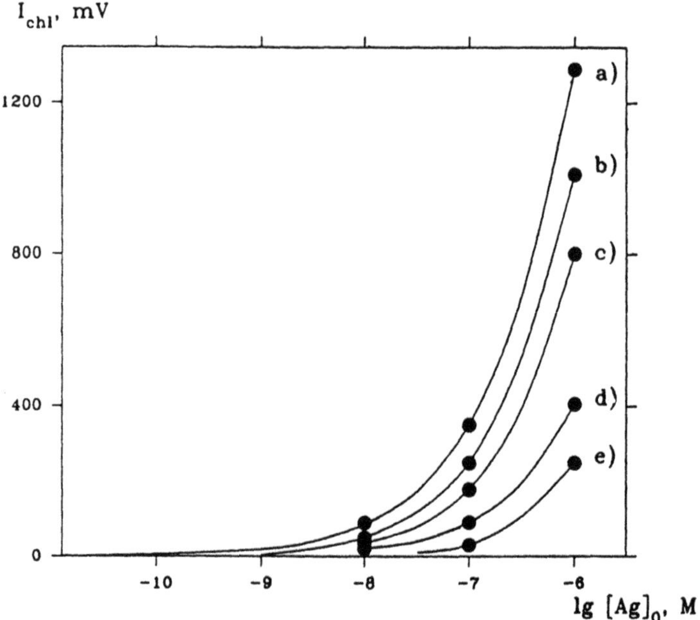

Figure 5. Standard curves of human IgG in a flow-injection analysis using quantitation of peroxidase label with an enhanced luminescent reaction. The incubation time for the first step of the assay varied as follows: (**a**) 10 min; (**b**) 7.5 min; (**c**) 5 min; (**d**) 2 min; (**e**) 1 min. The second step of the analysis was carried out with a 2-min incubation time. Initial concentration of antibody-enzyme conjugate was 1.0 x 10-7 M.

the immunochemical steps as well as label detection. The method developed is characterized by a total time of a single IgG determination of 10-15 min and a lower limit of detection of 10^{-9} M.

3.3 Flow-Injection Enzyme Immunoassay of T4

The third scheme of flow-injection enzyme immunoassay was developed using the example of determination of thyroxine [11].

Figure 3 illustrates the flow diagram of this assay. The antigen to be analyzed is sampled into the buffer stream and then interacts with a large excess of peroxidase-labeled antibodies to ensure fast and almost complete binding of the antigen. The unbound labeled antibodies are removed by passing through the affinity column containing the antigen immobilized on cyanogen bromide-activated Sepharose in a large excess with respect to the antibodies. The column showed no loss of activity after 100 experiments. After passing through the column, the solution contains the antibodies complexed to the antigen to be assayed.

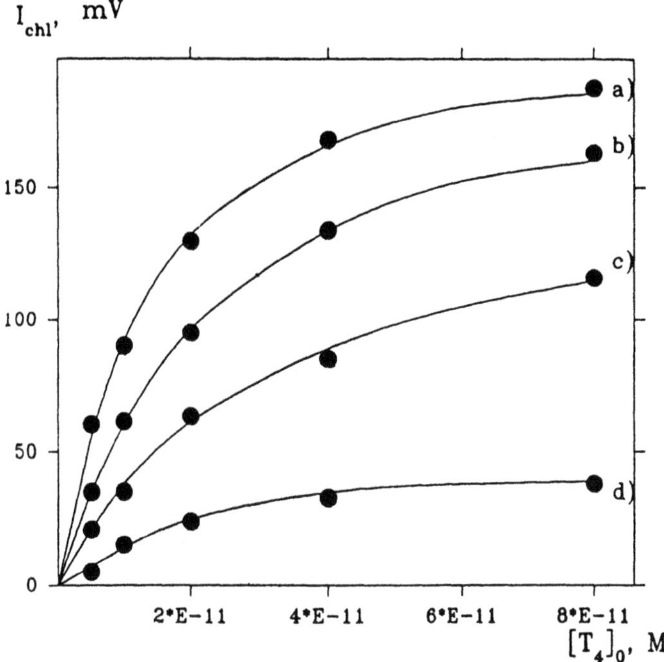

Figure 6. Standard curves for thyroxine obtained by the enhanced luminescent reaction. The incubation time for the first step of the assay was varied as follows: (**a**) 5 min; (**b**) 3 min; (**c**) 1.5 min; (**d**) 0.5 min. The incubation time was 1 min in the second step of the analysis. The initial concentration of antibody-enzyme conjugate was 3.0 x 10^{-9} M. The accumulation of the chemiluminescence signal was carried out for 3 min.

Computer simulation indicated that the sensitivity of the method at a fixed affinity of antibodies and concentration of labeled reagents is limited by the sensitivity of detection of the enzyme label. Used as a label, peroxidase was assayed by the enhanced chemiluminescence procedure using luminol, *p*-iodophenol and hydrogen peroxide.

Figure 6 shows the calibration graphs for assaying T4 in solution at different times of interaction between the haptens and the labeled antibodies. At sufficiently large time intervals (3 min or more) the lower detection limit of the antigen is about 6 x 10^{-11} M (curves a and b). When the exposure time is lower, the working range of assay moves to higher concentrations of the species measured. The time of separation of reagents on the affinity sorbent being 1 min in all cases.

The method described, in principle, allows the detection of both haptens and large antigens. In addition, it allows the measurement of small amounts of the antigen the determination of which by the non-competitive sandwich routine would take more than 3 h, and by the competitive monocentered version of the solid-state EIA is impossible. This is probably due both to the lack of diffusional restrictions, as the immunoreactions occur in solution, and to the utilization of a large molar excess of the labeled antibodies compared with the initial amount of the antigen analyzed.

The proposed methods of FIIA incorporating the flow-injection technique and the enhanced chemiluminescence detection is fast, reproducible, convenient for routine flow immunotesting, and is characterized by high sensitivity with respect to antigens. However, it is well known that the use of luminol chemiluminescence reaction for assay of serum samples in a homogeneous system leads to a high background, and current work is directed towards this problem.

3.4   Flow Immunosorbent Assay of 2,4-Dichlorophenoxyacetic Acid

Using a checkboard titration we found the best combination of antiserum and conjugate dilutions giving the highest sensitivity towards analyte. The calibration

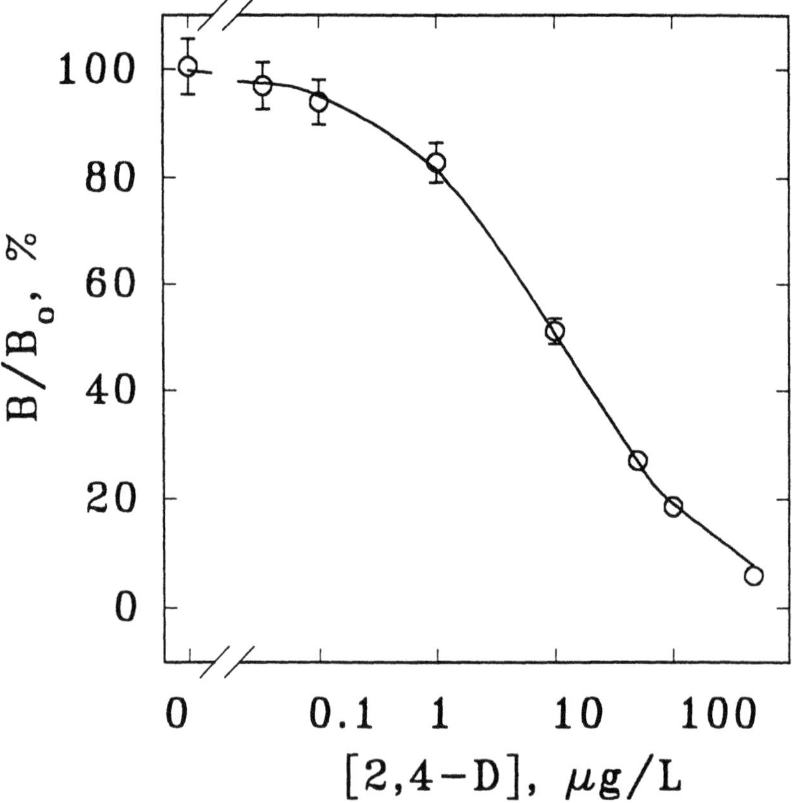

Figure 7.   Standard curve for the determination of 2,4-D using ELISA.

curves, constructed at optimal conditions are shown in Figure 7. To make a comparison easier, the curves are presented as B/B_0 graphs. The sensitivity (as concentration of hapten at $B/B_0 = 90\%$) of the assay was 1 µg/l. The maximum contaminant level for 2,4-D is 70 µg/l[12]. The sensitivity of the assay mainly depends on the quality of the antibodies and on the fine matching of the antibody-tracer pair. The reproducibility of the assay expressed as percent C.V. was in the range of 3 to 8%.

The performance characteristics of standard ELISA were then compared to those of affinity chromatography for the same antibody-tracer pair. The results of the assays with IC (as B/B_0) are shown in Figure 8. All steps of the antigen-antibody interactions in this case proceed in a kinetic, non-equilibrium mode. This allowed us to reduce incubation steps to a few minutes. The main reason for such rapid kinetics is that beads provide a lot of accessible surface area on which antigen-antibody interaction takes place. Antigen and tracer rapidly diffuse into the particles where the diffusion path is much shorter than in a well of a micro-plate. The full time of the assay was about 10 min per sample. All steps in the

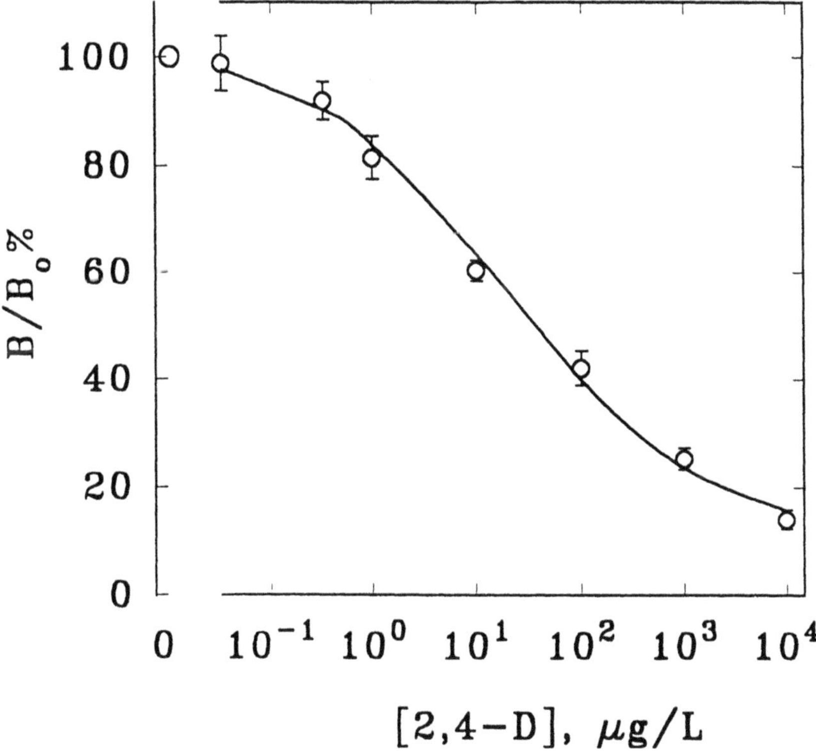

Figure 8. Standard curve for determination of 2,4-D with IC.

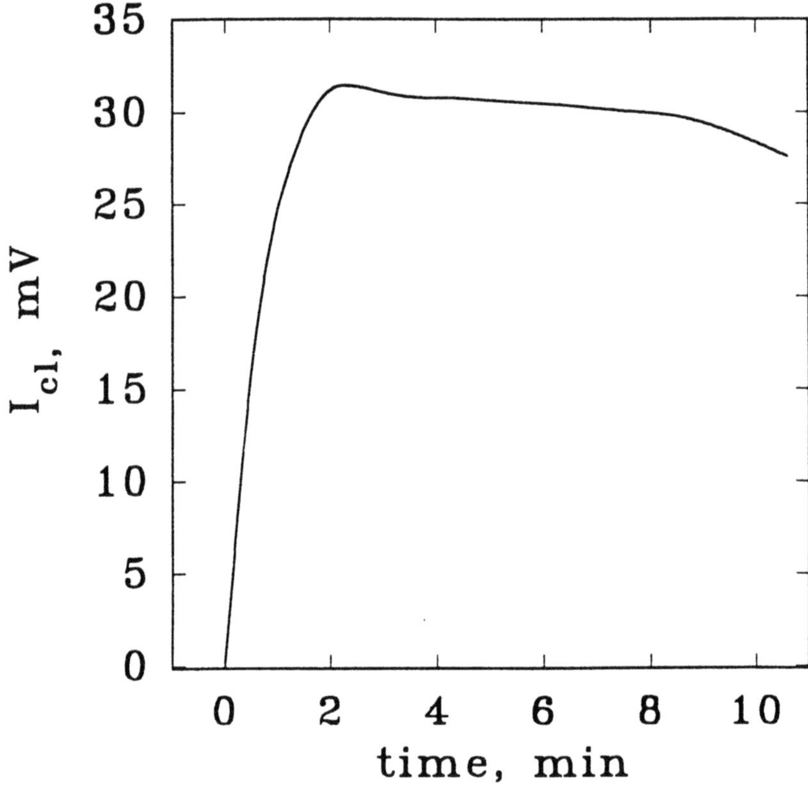

Figure 9.   Time-course of the reaction of enhanced chemiluminescence.

procedure, including detection, can be easily automated. In an automatic mode, several samples may be processed in parallel. This reduces the effective assay time per sample.

The assay could detect 1 µg/l. The coefficient of variation for all detection systems with affinity chromatography was in the range of 3 to 8%. Any non-equilibrium scheme of immunoassay suffers from relatively low precision (compared to an equilibrium one). Since the antigen-antibody reaction does not reach equilibrium, such procedures need careful time control. This requirement also limits the maximum number of samples which can be analyzed simulta-neously. The problem may be overcome using a microprocessor-controlled dispensing unit. Our estimation shows that up to 9 and 50 samples may be analyzed in parallel using manual and automatic dispensing, respectively.

The other possible sources of error are connected with column-to-column variation of sorption capacity of the gel. Our experiments show that these factors do not significantly contribute to the total error of the assay with ABICAPs. The kinetics of the signal is shown in Figure 9. The peak of the signal is observed in 1 to 2 min after addition of substrate mixture. Short detection time is of special importance for non-equilibrium types of immunoassay. Given the total time of all steps of antigen-antibody interaction of 5 min, a detection time of 5-15 min required by conventional methods becomes inappropriate. Such express techniques may benefit from switching to "faster" detection systems like enhanced chemiluminescence.

In IC we used the same volume and dilution of antibodies and conjugate as in ELISA. The flow-through construction of the main unit for IC allows one to vary sample volume over a wide range. This option can be used to adapt the sensitivity of assay to the concentration of analyte. When the concentration is low, a large volume of sample can be pushed through the column to concentrate the analyte. Conversely, concentrated samples may be applied onto the column in a very small volume, down to 1 μl. This option enables one to avoid a dilution step, which is one of significant sources of error in ELISA.

Previously, IC was mainly used with the sandwich scheme of immunoassay [13-15]. In that case, a large excess of antibodies over analyte helps rapid capture. We showed the possibility of applying the IC technique to a competitive type of assay where the number of binding sites is limited.

In this work an IC method to analyze 2,4-D was developed. The sensitivity of assays was the same and precision a little lower than with ELISA. The total assay time per sample was 10 min and may be further reduced in an automatic mode. The procedure can be easily adopted for quasi-continuous control of pesticide contamination. IC may well be used as express assay in out-of-laboratory conditions.

ABBREVIATIONS

EIA, enzyme immunoassay; FIA, flow injection analysis; FIIA, flow injection immunoanalysis; ABICAP, Antibody Immunocolumn for Analytical Processes; PBS, phosphate-buffer saline; PBST, PBS supplemented with 0.05% Triton X-100; TMB, 3,3',5,5'-tetramethylbenzidine; HRP, horseradish peroxidase; T4, thyroxine; BSA, bovine serum albumine; 2,4-D, 2,4-dichlorophenoxyacetic acid; C.V., coefficient of variation; IC, immunochromatography.

REFERENCES

[1]   Egorov, A.M., Osipov A.P., Dzantiev, B.B., Gavrilova, E.M. (1991). *Theory and practice of enzyme immunoassay*, pp. 291. *High School Press, Moscow.*

[2]   Dzantiev, B.B., Osipov, A.P. (1987). Classification and characteristics of enzyme immunoassay methods. In: Egorov, A.M.(ed.) *Itogi Nauki i Tekhniki*, (*Serie Biotechnology*, (Egorov, A.M., Ed.), Vol.3, pp. 3-85. VINITI Press, Moscow.

[3]   Arefyev, A.A., Osipov, A.P., Egorov, A.M. (1987). Enzyme-linked immunoassay of human IgG in kinetic mode. *J. Microbiol. Epidem. Immunobiol.* (Russian), **9**, 27-32.

[4]   Osipov, A.P., Arefyev, A.A., Yegorov, A.M. (1989). Flow-injection immunoenzyme assay of human IgG in the kinetic regime. *Mendeleev Chemistry Journal* (Zhurnal Vses. Khim. Ob-va im. D.I.Mendeleeva, Russian), **34**;119-120.

[5]   Ruzicka, J. (1981). *Flow Injection Analysis*, pp.207. John Wiley and Sons, New York.

[6]   Whitehead, T.P., Thorpe, G.H.G., Carter, T.J.N., Groucutt, C., Kricka, L.J. (1983). Enhanced luminescence procedure for sensitive determination of peroxidase-labelled conjugates in immunoassay. *Nature*, **305**,158-159.

[7]   Vlasenko, S.B., Arefyev, A.A., Osipov, A.P., Klimov, A.D., Kim, B.B., Gorovitz, E.L., Gavrilova, E.M., Egorov, A.M. (1989). An investigation on catalytic mechanism of enhanced chemiluminescence: immunochemical applications of this reaction. *Biolum Chemilum.* **4**, 164-176.

[8]   Hall, J.C., Deschamps, R.J.A., Krieg, K.K. (1989). Immunoassay for the detection of 2,4-D and picogram in river water and urine. *J.Agric.Food Chem.*, **37**, 981-984.

[9]   Nakane, P.K., and Kawaoi, A. (1974). Peroxidase-labeled antibody. A new method of conjugation. *J.Histochem.Cytochem.*, **22**, 1064-1091.

[10] Osipov, A.P., Arefyev, A.A., Vlasenko, S.B., Gavrilova, E.M., Egorov, A.M. (1989). Flow-injection enzyme immunoassay for human IgG by using enhanced chemiluminescence reaction for horseradish peroxidase label quantitation. *Anal. Lett.*, **22**, 1841-1859.

[11] Arefyev, A.A., Eremin, S.A., Vlasenko, S.B., Osipov, A.P., Egorov, A.M. (1990). Flow-injection enzyme immunoassay of haptens with enhanced chemiluminescence detection. *Anal.Chim.Acta*, **237**, 285-289.

[12] U.S. EPA, (1989) *Drinking Water Health Advisory*: Pesticides. Lewis Publishers, Chelsea, MI.

[13] Hage, D.S., Kao, P.C. (1991). High-perfomance immunoaffinity chromatography and chemiluminescent detection in the automation of a parathyroid hormone sandwich immunoassay. *Anal.Chem.*, **63**, 586-595.

[14] Afeyan, N.B., Gordon, N.F., Regnier, F.E. (1992). Automated real time immunoassay of biomolecules. *Nature*, **358**, 603-604.

[15] Stocklein, W., Jager, V. Schmid R.D. (1991). Monitoring of mouse immunoglobulin G by flow-injection analytical affinity chromatography. *Anal. Chim. Acta*, **245**, 1-6.

THE BACKGROUND FOR CREATING BIOSENSORS BASED ON NUCLEIC ACID MOLECULES

Y. M. Yevdokimov, S. G. Skuridin, and B. A. Chernuha

OUTLINE

ABSTRACT

The main theoretical principles used for creating classical and disposable biosensing units based on linear single- and double-stranded nucleic acid molecules are considered in this paper. These units are capable of realizing different modes of "recognition" of biologically important substances—from nucleotide sequences

Advances in Biosensors
Volume 3, pages 143-164.
Copyright © 1995 by JAI Press Inc.

ISBN:1-55938-535-9

forming complementary hybrid complexes to a wide range of antitumor compounds reacting with nitrogen bases of nucleic acids. Some problems so far unsolved in this area are outlined.

1. INTRODUCTION

Theoretically speaking, any biopolymeric molecule can be used to create biosensing units for biosensors. Yet, as can be seen from the reviews in this volume as well as from literature, most biosensors are based on enzymes (proteins). The application of other biopolymers such as nucleic acids to the creation of biosensors is described in only a few publications.

It is important to appreciate why nucleic acids form a sound foundation for creating biosensing units. The answer is predetermined by a combination of several factors. First, single- and double-stranded desoxyribonucleic acid (DNA) molecules seem to be the most important of all biological molecules: they represent the genetic material of living cells.

Second, many chemical or biological compounds present in the environment can interact with the genetic material of living cells. Hence, the detection of biological or chemical compounds that influence genetic material in living cells is relevant. Third during the life cycle of a cell or infection nucleic acid fragments can appear that carry incorrect genetic information, therefore the estimation of these sequences is necessary (genetic screening). In addition, pharmaceutical manufacturers need to quantify very small amounts of contaminating nucleic acids in recombinant proteins and monoclonal antibodies intended for therapeutic use. Finally, the physical chemistry of isolated nucleic acid molecules is well-studied.

The following characteristics of the DNA molecules should be considered as important in the design of biosensors:

1. Right-handed double-stranded nucleic acids exist in different spatial forms (A, B and Z, linear, circular, superhelical, etc.) that vary noticeably in their physicochemical characteristics. A transition between these forms is initiated by changes in the surrounding conditions.
2. Double-stranded molecules of nucleic acids whose negatively charged phosphate groups are neutralized with counterions are "rigid"; single-stranded nucleic acids are "flexible."
3. Nucleic acid molecules are optically active.
4. Each chain of a nucleic acid molecule contains nitrogen bases (chromophores absorbing in the UV spectral region) that form complementary hydrogen bonds with those in the other chain (complementary "recognition").
5. Each chain of a nucleic acid molecule contains many chemical groups which differ in their reactivity.

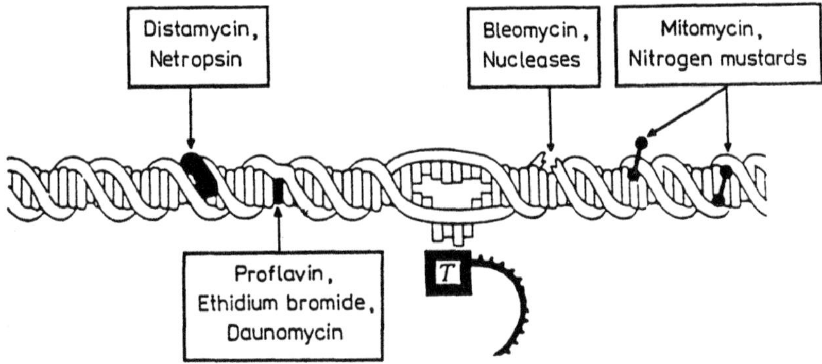

Figure 1. Modes of structural "addressing" of biologically active compounds by double-stranded nucleic acid molecules.

6. The helical structure of double-stranded nucleic acids in combination with the specific distribution of reactive groups on the surface of these molecules can "address" those compounds that interact with nitrogen bases of nucleic acids.

7. Different biologically active compounds can "recognize" different chemical groups in the content of DNA (Figure 1). The scheme shows that biologically active molecules (antibiotics, antitumor compounds, enzymes, etc.), which differ in their structures, can, according to above items 6 and 7, "recognize" different elements in the DNA structure. The "recognition" results in structural changes (i.e. an increase of the distance between nitrogen bases, alteration of their inclination angle, formation of "cross-links" between bases, splitting of the DNA sugar-phosphate chain(s), etc.).

Therefore, in contrast to proteins (enzymes), nucleic acid molecules have several specific properties. A combination of these properties opens up a possibility for using different "recognition" principles for detection of compounds which interact with nucleic acids in this or that way.

BIOSENSORS BASED ON SINGLE-STRANDED NUCLEIC ACIDS

Problems arising when constructing biosensors based on single-stranded nucleic acid or synthetic polynucleotide molecules are reviewed in [refs. 1-3]. We discuss here only several principal questions pertinent to creation of these biosensors.

The creation of biosensing units based on single- stranded nucleic acids involves several steps (Figure 2). First, a single-stranded nucleic acid fragment

1. Scheme of biosensing unit based on single-stranded nucleic acid molecules.

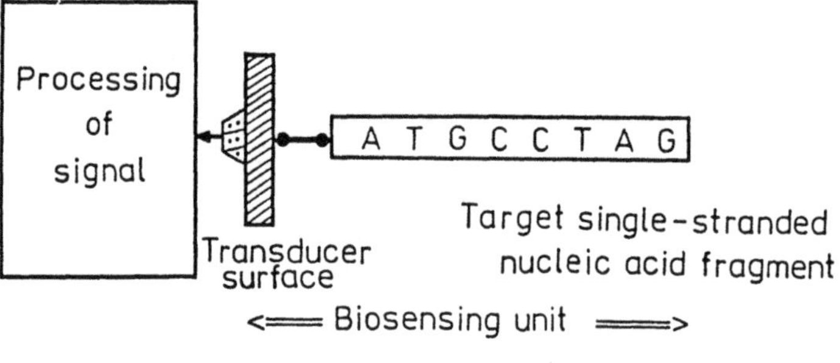

2. Probe.

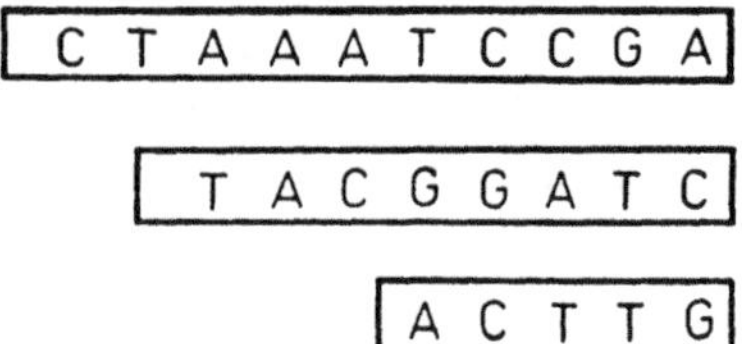

3. Analysis of probe.

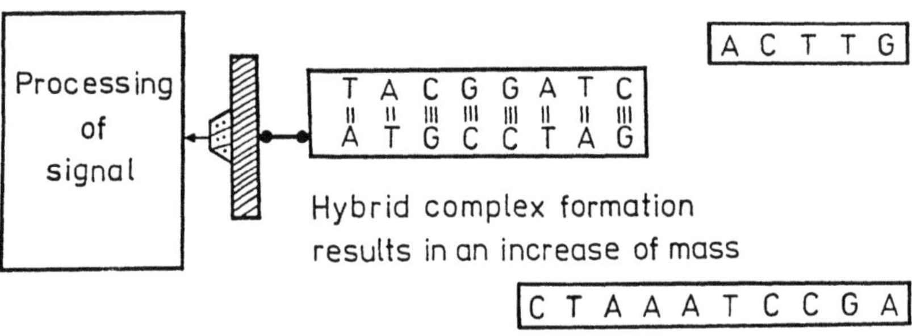

Figure 2. A scheme for the biosensing unit based on single-stranded nucleic acid molecules.

containing a sequence of nitrogen bases which is of interest for the researcher (a "target" nucleic acid) has to be either isolated from cells or synthesized. The choice of this sequence is of medical importance because it is associated with a definite molecular-genetic disease or bacterial (viral) infection. Second, the molecules of the "target" should be linked onto a solid support (the "carrier"), i.e., these molecules are immobilized because of chemical binding to the "carrier." Nylon or nitrocellulose films are used most often [2]. Immobilization should be performed by a chemical reaction which permits a high density of molecules to be attained on the "carrier" with retention of their initial chemical reactivity.

The fragment of nucleic acid molecules ("target") linked to the "carrier" represents a biosensing unit.

If the biosensing unit is placed into the medium to be analyzed, which contains foreign single-stranded nucleic acids differing in base sequences ("probes"), there may occur a "recognition" of only those base pairs in the content of alien nucleic acids which are complementary to the base sequence of the "target." "Recognition" is accompanied under specific conditions (ionic strength, pH, length of nucleic acid fragments, etc.) with formation of a complementary complex. Thus, a biosensing unit can "seek out" only those DNA molecules from their mixture which bear responsibility for a definite genetic disease. This process is often referred to as a hybridization; under definite experimental conditions a hybrid complex can dissociate, hence the biosensing unit is restored for a new test. The complex formation between the "target" and the "probe" results in the change of several physicochemical parameters of a "target," such as mass, absorption or diffusion coefficients, and capacity of layer formed on transducer surface. Theoretically, these changes can be used for detection of a presence of the "probe" in the hybrid complex. The most evident practical realization of biosensors based on single-stranded nucleic acid molecules is detection of the mass changes resulting from complementary hybrid formation. As an example of a practical use of mass changes a detection of some synthetic oligonucleotide fragments by piezo-electric method [4,5] is considered. The concept of this method is clear. It is based on the Sauerbrey Equation [4], which describes the frequency-to-mass relationship,

$$\Delta f = -2.3 \times 10^6 \times F^2 \times \Delta M / A \qquad (1)$$

where: Δf is the change in fundamental frequency of piezoelectric crystal coated with biopolymeric molecules (Hz); F is the resonant frequency of the crystal (Hz); ΔM is the mass of the deposited coating (g); A - is the coated area (cm^2).

If a layer of biopolymeric molecules on the piezoelectric surface can be created, any mass changes of the device (e.g., after formation of a complementary complex) result in alteration of the resonant frequency, which can be detected at very high sensitivity. (The theoretical detection limit is estimated to be about 10^{-12} g.)

In practice [6], fragments of synthetic nucleic acids (poly C, poly U, etc.) were covalently bound to a copolymer (styrene-acrylic acid)-modified surface of quartz. When these immobilized "targets" were melted and then incubated with complementary "probes" in solution, the complex formation between the "target" and the "probe" resulted in an increase of the effective mass of the crystal. After washing away indifferently bound material, the mass increase was detected as a decrease of several hundred Hz in the crystal's resonant frequency.

Piezoelectric detection is of interest to clinicians, as it could be applied to identification of different viruses and bacteria.

It should be added that this method has certain limitations because the oscillation of the quartz is strongly damped by contaminants arising from absorption of impurities as well as from binding of the supporting polymeric film to the quartz. If too much weight is deposited on the crystal, the vibration goes into overtones and the frequency changes no longer have the linear relationship predicted by the Sauerbrey Equation.

A further possibility for creating biosensors is based on detection of a special "label" which is incorporated into the content of the "probe" before hybrid complex formation. Here we are dealing with creating biosensors of the disposable type.

Various substances have been either used or advocated as labels for nucleic acid "probes" and these are summarized in ref. 2. An ideal label for a DNA would have the following properties:

1. It should readily be linked to chemical groups of nucleic acids.
2. Low concentrations should be detected by simple instruments.
3. Its properties should change noticeably when a hybrid complex is formed (thus facilitating detection of nonspecific DNA complexes).
4. The label must be stable at elevated temperatures that are often needed for hybridization.

Two main strategies to incorporate a label into nucleic acid probes have been evolved:

1. "Direct" labeling, in which a label is linked directly via a covalent binding to chemical groups of probes.
2. "Indirect" labeling, in which either hapten is linked to the chemical groups of the probe and detected by using a labelled binding protein with specificity for this hapten.

Determination of foreign nucleic acid in hybrid complexes by a radioactive label (e.g. ^{32}P) introduced into the probe is widely applied. This is preconditioned by the existence of very sensitive radioactive counters and a well-established technique for detection of the label.

In the simplest case, the biosensing unit operation resides in the fact that the nitrogen bases of the single-stranded nucleic acid "recognize" the complementary fragment labeled by ^{32}P (Figure 3). The appearance of the label in the content of the hybrid complex points to the presence of foreign nucleic acid in the analyte. A subsequent removal of the biosensing unit from contact with the analyte and

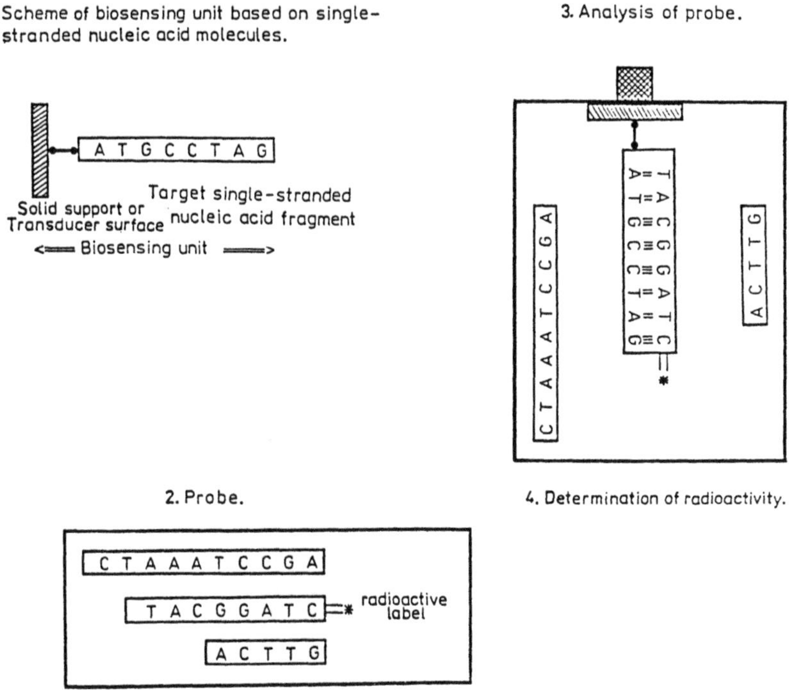

1. Scheme of biosensing unit based on single-stranded nucleic acid molecules.

3. Analysis of probe.

A T G C C T A G

Target single-stranded nucleic acid fragment

Solid support or Transducer surface

<=== Biosensing unit ===>

2. Probe.

C T A A A T C C G A

T A C G G A T C —* radioactive label

A C T T G

4. Determination of radioactivity.

Figure 3. A disposable biosensing unit based on single-stranded nucleic acid molecules.

determination of radioactivity allows the concentration of the DNA probe to be estimated. According to published data, this strategy makes it possible to determine about 10 picograms of a single-stranded DNA [7].

The above listed biosensors, known as "DNA probes" [2], are used to detect the presence of nucleic acid sequences associated with genetic diseases (e.g. sickle cell anemia, phenylketonuria) or analysis of restriction fragment length polymorphism (e.g. adult polycystic kidney disease) [7]. Various aspects which involve biosensors based on single-stranded DNA's have been discussed in relation to medical problems [1,2,7].

Considering the technical aspects of these biosensors, it should be pointed out that application of radioactive labeled compounds is limited by several factors. First, the most commonly used label (^{32}P) is expensive and has a short half-life time ($t_{1/2}$ = 14.3 days) necessitating frequent preparation of probes. Second, the application of labeled compounds is hazardous. Third, the quantitative analysis can be made only in laboratories especially equipped for that purpose. Besides, from Figure 3 it can be seen that in this method contamination by unexpected sequences may be missed altogether.

Since the application of radioactive labels is sometimes difficult, attempts are being made to create new methods for detecting foreign DNA using alternative techniques whose sensitivity is not inferior to that of a radioactive assay, such as electrochemical or optical, in combination with enzymatic "amplification." A new light-addressed potentiometric sensor (LAPS) proposed by Molecular Devices [8] can pick up as little as 2 picograms of DNA, including fragments as short as 200 base pairs.

Thus, it can be stated that elaboration of the methods for detecting a the signal generated by biosensing units based on single-stranded nucleic acid molecules is progressing. Construction of biosensors based on single-stranded nucleic acids is a field of science with a very narrow gap between the theoretical substantiation and its practical application.

3. BIOSENSING UNITS BASED ON LIQUID–CRYSTALLINE DISPERSIONS OF LINEAR NUCLEIC ACID MOLECULES

The ability of linear double-stranded nucleic acid molecules to "recognize" different chemical or biologically active compounds is used with the highest efficiency in biosensing units based on lyotropic liquid-crystalline dispersions formed from these molecules. Different versions of "recognition" can be realized in such biosensing units.

The creation of double-stranded nucleic acid-based biosensing units makes use of experimental data specific to liquid crystals as well as to liquid-crystalline dispersions of nucleic acids—in particular, of double-stranded DNA.

Since these concepts have only recently been formulated [9,10], we shall illustrate merely the principal aspects of this problem.

It is well known that the right-handed rigid optically active nucleic acid molecules (mol. wt. $< 1 \times 10^6$) at a definite concentration in solution ($C_{NA} > C_{NA}{}^{cr}$) tend, because of phase separation, to form lyotropic cholesteric liquid crystals or liquid-crystalline dispersions. The distance between adjacent DNA molecules in the liquid crystals is as large as 25 to 40 A° [11]. The ordered arrangement of linear double-stranded DNA molecules does not influence their structural peculiarities; thus these molecules in liquid crystals belong to the B-family.

Double-stranded nucleic acid cholesteric liquid crystals are, in essence, "dyed" cholesterics because they contain "chromophores" (nitrogen bases) that absorb in the UV region of the spectrum. An abnormal optical activity in the absorption region of the chromophores is specific to "dyed" cholesterics [12-14]. Such optical activity may be expressed as an intense (abnormal) band in the circular dichroism (CD) or optical rotation (OR) spectra. Figure 4 compares the CD spectra specific of initial linear double-stranded DNA molecules (B-form of DNA), those of the DNA liquid-crystalline dispersion, and of the DNA liquid crystals [15]. It proves that for the DNA liquid crystals in the absorption region of nitrogen bases ($\lambda \sim 260$

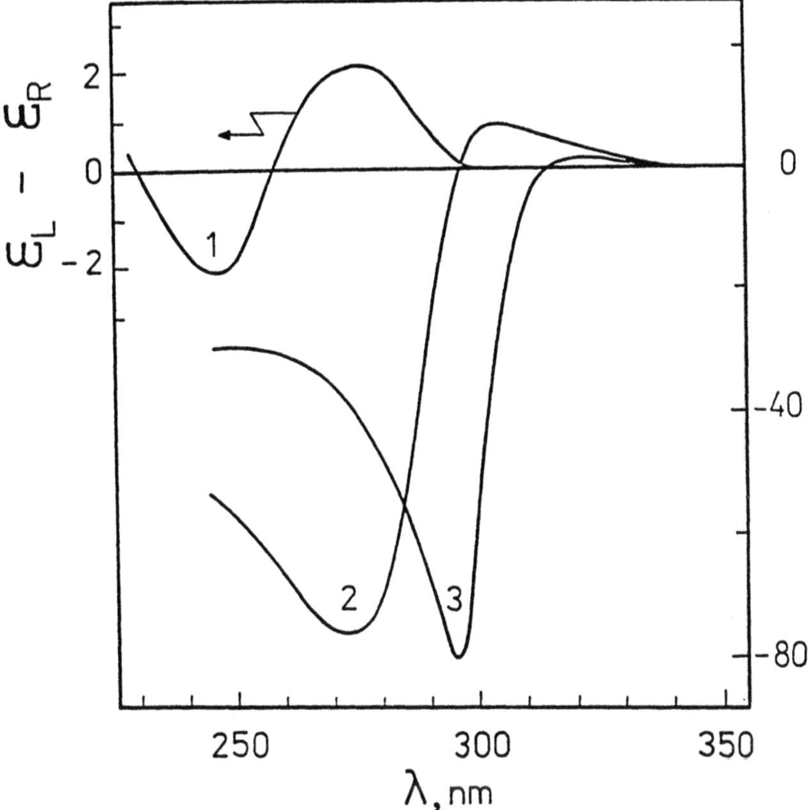

Figure 4. The CD spectra of linear B-form DNA (*curve 1*), its cholesteric liquid-crystalline dispersion formed in poly(ethyleneglycol)-containing (PEG) solution (*curve 2*) and thin layer (~ 20 μm) of DNA cholesteric liquid crystals (*curve 3*). (1) 0.3 M NaCl + 10^{-2} M phosphate buffer; (**2, 3**) 0.3 M NaCl + 10^{-2} M phosphate buffer, C_{PEG} = 170 mg ml-1, mol. mass of PEG = 4000.

nm) an intense negative band appears, its amplitude exceeding dozens of times the amplitude of the band typical of the initial DNA molecules.

In the frame of modern theory [16,17] the dependence of the amplitude of the intense band in the CD spectra specific to a cholesteric liquid crystal film containing added chromophores ("dyed" cholesterics) was described [12-14,16,17] in terms of the known parameters,

$$\Delta A_{\upsilon_i} \sim N \times f_j(\alpha), f_k(\varepsilon_{||}/\varepsilon_m ; \varepsilon_{\perp}/\varepsilon_m), f_l(\upsilon_i), f_m(T) \times M \qquad (2)$$

where: $\Delta A_{\upsilon_i} = \Delta A_L - \Delta A_R$ is the circular dichroism of the film at frequency υ_i; α is the inclination angle of chromophores with respect to the director of quasine-matic layer; ε_m is the dielectric constant of a medium; $\varepsilon_{||}$, $\varepsilon_{\perp}$ are the longitudinal and transverse components of dielectric permeability of the polymeric molecule; T is temperature; M, N are the arbitrary constants.

After introduction of the practically detectable parameters specific of the DNA nitrogen bases into expression (2) the negative sign of the intense band in the CD spectrum of the DNA liquid crystals (Figure 4) shows that a left-handed cholesteric helix is formed from linear double-stranded right-handed (B-form) DNA molecules.

Expression (2) reflects the role of different factors influencing properties of polymeric liquid crystals.

There are certain limitations, however, for using the film formed from DNA liquid crystals in practice: (1) a low stability of these films, and (2) high concentration of the DNA in the liquid-crystal phase, which can influence the diffusion-controlled process of interaction of chemicals with the nitrogen bases. On the other hand, many experimental data [11] show that the main factors which govern the properties of the DNA liquid crystals, play the same role for dispersions. For example, a comparison of the CD spectra for the DNA liquid crystals and their liquid-crystalline dispersions (Figure 4) demonstrates that the abnormal optical activity which is specific to the cholesteric DNA liquid crystals is also specific to liquid-crystalline dispersions. Differences between curves 2 and 3 are explained in terms of the theory [18] that describes optical properties of liquid crystals and dispersions of different chemical compounds. In terms of theory [19], a so-called "size effect" induces more "flexibility" into the properties of the DNA liquid-crystalline dispersions in comparison with the properties of the DNA liquid crystals. Therefore, the use of DNA liquid-crystalline dispersions allows the influence of different factors to be detected more easily. These facts, taken together, can open a possibility for practical use of DNA liquid-crystalline dispersions as biosensing units of the disposable type.

Taking into account that the exact equation that describes optical properties of DNA liquid-crystalline dispersions is absent, that the pitch for DNA liquid crystals [11] is much larger than λ_{max} of absorption of nitrogen bases (2 μm $>>$ 0.2 μm), that expression (2) does not specify the size of a liquid-crystalline phase, neglecting the light-scattering as well as nonhomogeneous distribution of particles of the liquid-crystalline dispersion in a polymer-containing solvent, we can rewrite expression (2) in a following form,

$$\Delta A_{\upsilon_i} \sim N \times f(C_{NA}), f_j(\alpha), f_k(\varepsilon_{||}/\varepsilon_m \,; \varepsilon_{\perp}/\varepsilon_m\,), f_l(\upsilon_i\,), f_m(T) \times M \qquad (3)$$

where: ΔA_{υ_i} is the effective amplitude of the intense band in the CD spectrum observed for dispersion in the region of absorption of nitrogen bases; C_{NA} is the

DNA concentration used for formation of the liquid-crystalline dispersions; all other parameters are defined above.

Because formation of the DNA liquid-crystalline dispersions takes place under fixed experimental conditions (i.e., C_{NA}, T, ε_m are constants) and because the mode of orientation of the DNA nitrogen bases is practically unchanged under any conditions as long as DNA molecules are not denatured (i.e., $\varepsilon_{||}$, $\varepsilon_{\perp}$, and α are constants, too), expression (3) is reduced to (4):

$$\Delta A_{\upsilon_i} \sim f(C_{NA}) \times B \qquad\qquad (4)$$

Therefore, the amplitude of the band in the CD spectra of the DNA cholesteric liquid-crystalline dispersions can be proportional to the DNA concentration.

Figure 5 illustrates the dependence of the ΔA_{270} value measured in the CD spectra of DNA liquid-crystalline dispersions which are formed because of the phase separation in the polymer-containing solution [20] versus concentration of the DNA used for preparation of these dispersions. The constant value of $\Delta\varepsilon_{270}$

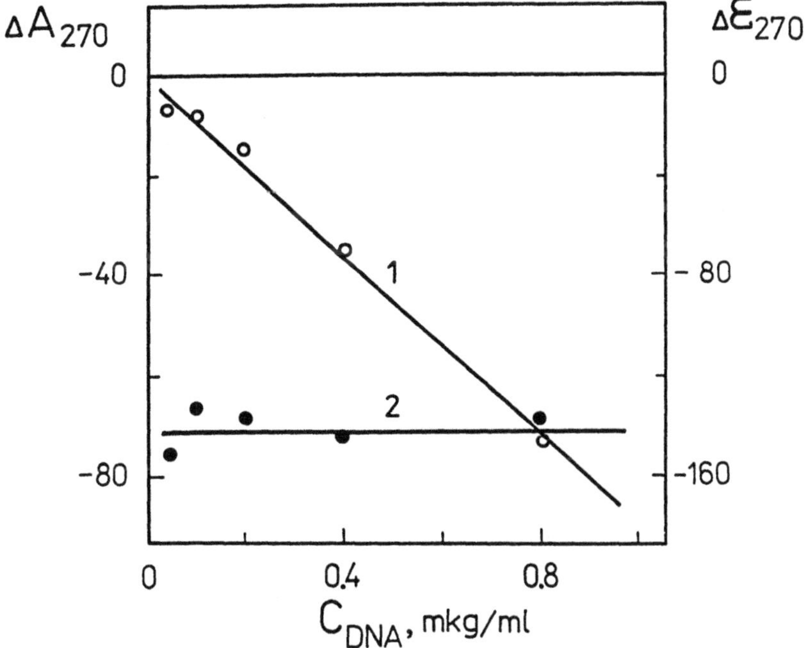

Figure 5. The dependence of the amplitude of the negative band in the CD spectra of the DNA liquid-crystalline dispersions at $\lambda = 270$ nm, (ΔA_{270}, (*line 1*) and $\Delta\varepsilon_{270}$ value (*line 2*) versus DNA concentration. 0.3 M NaCl + 10^{-2} M phosphate buffer, C_{PEG} = 180 mg ml^{-1}, mol. mass of PEG = = 4000, ΔA_{270} in cm, 1 cm = 5 x 10^{-6} optical

($\Delta\varepsilon_{270} = \Delta A_{270}/C_{NA}$, at $l = 1$ cm, where l is the length of optical cell) shows clearly that at definite experimental conditions the ΔA_{270} value is directly proportional to the DNA concentration. Strictly speaking, one can expect a deviation of this dependence from *linearity* both at a very low DNA concentration, because formation of liquid-crystalline dispersions is limited to the number of molecules forming one particle of the liquid-crystalline dispersion ($\sim 10^4$ DNA molecules [21]), and at a high DNA concentration, because here nonspecific aggregation influences the optical properties of DNA liquid-crystalline dispersions.

It follows from expression (3) that if the characteristics of the solvent (T, ε_m) are kept constant and the properties of the DNA molecules varied (i.e., the ratio $\varepsilon_{||}/\varepsilon_{\perp}$ is changed), one can create liquid-crystalline dispersions differing in their abnormal optical activity. The alteration, for example, of the physico-chemical properties of the DNA nitrogen bases, because of their modification, will change the amplitude of the band in the CD spectrum specific to the original DNA liquid-crystalline dispersions in spite of the sign of this band. It means that the intense band in the CD spectrum permits the mode of spatial orientation of the DNA nitrogen bases to be visualized.

Therefore, the amplitude of the intense band in the CD spectrum of the DNA liquid-crystalline dispersions represents a criterion, which can reflect different factors influencing the DNA structure, i.e., genetic material.

A few examples that illustrate the analytical "possibilities" of the DNA liquid-crystalline dispersions are given below.

It was shown [22] that UV irradiation of DNA molecules in the presence of sodium hypophosphite (NaH_2PO_2) is accompanied by the following chemical reaction:

Differences between the normal nitrogen base—thymine (compound A) and dihydrothymine (compound B)—consist of alteration of the electronic structure and in breaking of its planar character. Figure 6B compares time dependences of the amplitude of the negative band in the CD spectra obtained for DNA liquid-crystalline dispersions formed under different conditions of UV irradiation. Figure 6 shows that the above very fine irreversible changes in the structure of DNA thymine residues strongly alter the cholesteric organization of DNA liquid-crystalline dispersions.

Another example of chemical modification of DNA molecules is formation of a complex between nitrogen bases and the antitumor compound [23]. Figure 7A compares the CD spectrum of native, initial, linear form of DNA with those of liquid-crystalline dispersions formed from native and from modified by antitumor

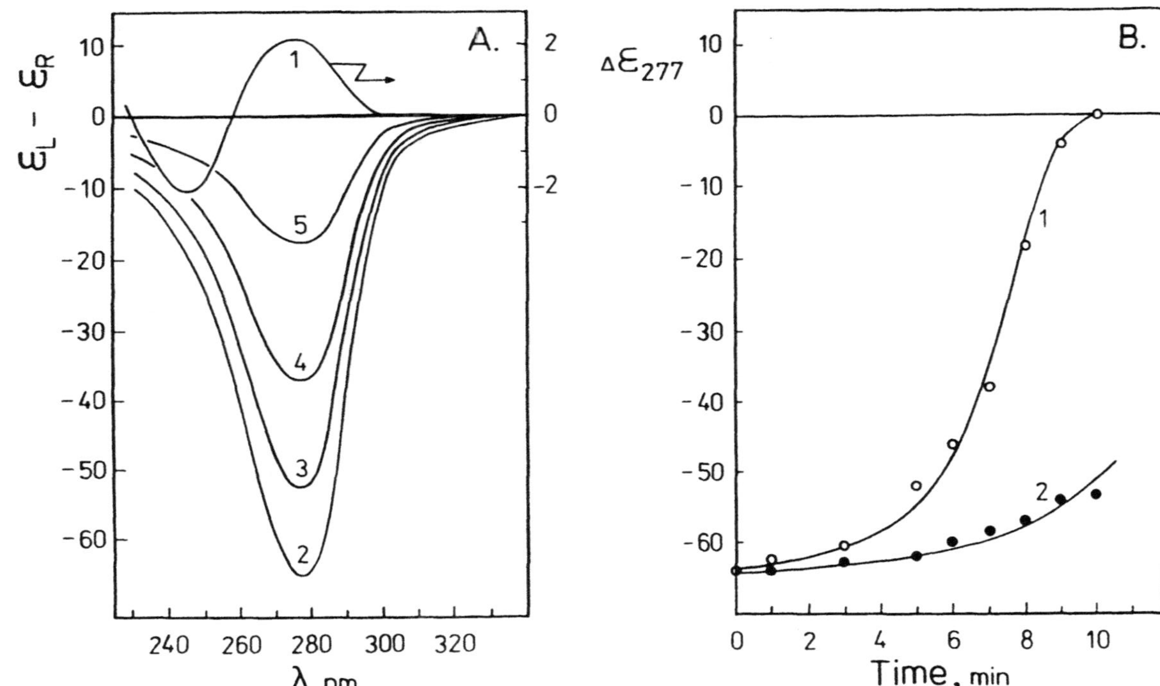

Figure 6. (**A**) The CD spectra of liquid-crystalline dispersions formed from initial (*curve 2*) and UV irradiated ($\lambda = 254$ nm) in 2 M NaH_2PO_2 solution DNA molecules (*curves 3-5*). The time of UV treatment (min): 2.0, 3.5, 4.7, 5.8. $C_{PEG} = 170$ mg ml^{-1}, mol. mass of PEG = 4000. The CD spectrum of linear B-form of DNA (*curve 1*). (**B**). The dependence of the amplitude of the negative band in the CD spectra ($\lambda = 277$ nm) of liquid-crystalline dispersions formed from DNA molecules after UV treatment ($\lambda = 254$ nm) in the presence of 2 M NaH_2PO_2 (*curve 1*) and in its absence (*curve 2*) versus time of UV treatment. $C_{PEG} = 170$ mg ml^{-1}, mol. mass of PEG = 4000.

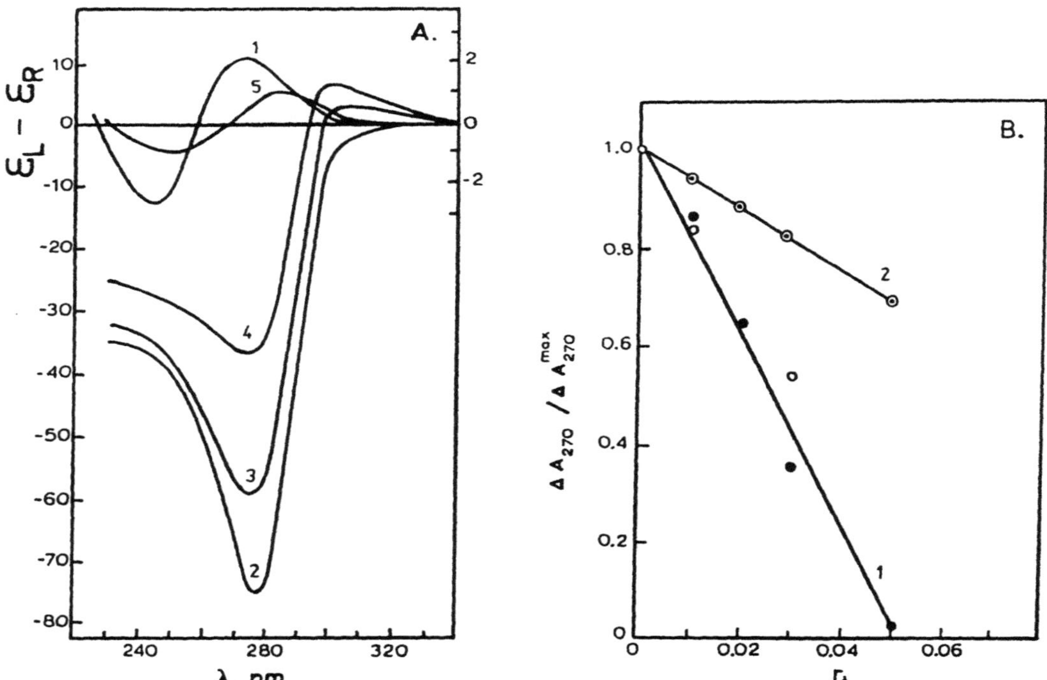

Figure 7. **(A)** The CD spectra of liquid-crystalline dispersions formed from initial DNA (*curve 2*) and DNA modified with DDP (*curves 3-5*). (2) $r_t = 0$, (3) $r_t = 0.01$, (4) $r_t = 0.03$, (5) $r_t = 0.1$. 0.5 M $NaClO_4$, $C_{PEG} = 170$ mg ml^{-1}, mol. mass of PEG = 4000. The r_t value is the concentration ratio (mol/mol) of DDP added to solution / nucleotide. (1) The CD spectrum of linear B-form DNA. 0.5 M $NaClO_4$. (2-4) Left ordinate; (1, 5) right ordinate. **(B)** The dependence of the relative amplitude of the negative band in the CD spectrum ($\lambda = 270$ nm) of liquid-crystalline dispersions prepared from DDP-modified DNA (*line 1*) and after their treatment with NaCN (*line 2*) versus r_t value.

compound, *cis*-diamminedichlorplatinum(II) (DDP). Attention is drawn to a few phenomena. First, according to the above concept, the higher the level of the DNA modification the lower the amplitude of the band in the CD spectra. It means, that covalent binding of DDP to N7 atoms of neighboring purine residues (mainly guanines) perturbs the orientation of adjacent nitrogen bases. Thus, the properties of modified DNA liquid-crystalline dispersions begin to differ from those corresponding to cholesterics. Second, at definite degrees of modification of nitrogen bases there is a linear dependence between the concentration of DDP which induces both the alteration of optical properties for the DNA liquid-crystalline dispersions and the amplitude of the negative band (Figure 7B). It means that there is a region of concentration where one can estimate the presence of DDP with good accuracy. Finally, the alteration of the properties of DNA liquid-crystalline dispersions induced by DDP interaction is reversible—addition of NaCN destroying the complex between DDP and nitrogen bases restores ~ 70 % of initial optical activity.

Therefore, the intense negative band in the CD spectrum specific to cholesteric liquid-crystalline dispersions of DNA molecules can reflect both the details of the DNA structure (base sequence, alteration of the structure, etc.) and those of the media. Under fixed experimental conditions it can be used as a criterion for detection of alteration of properties of the linear double-stranded DNA molecules under different conditions.

Above we have considered examples when compounds reacting with the DNA nitrogen bases were "colorless," i.e., they do not possess absorption at all or their absorption bands were located in the UV region. Detection of colored biologically active compounds which possess absorption in the region of the spectrum far from the DNA absorption can be performed in a more simple way. This approach is based on the X-ray investigation of liquid crystals formed from DNA molecules. Because the distance between adjacent DNA molecules in the cholesteric liquid crystal phase is equal to ~ 40 A° [11], the ordering of nucleic acid molecules in liquid crystals as well as liquid-crystalline dispersions does not destroy their ability to "recognize" and "address" the chemical compounds that react with the nitrogen bases.

For colored compounds forming strong complexes with DNA, i.e. for compounds that are fixed on a DNA molecule, the same rules apply that define the appearance of abnormal optical activity for nitrogen bases rigidly fixed in the structure of the DNA molecule. But, expression (3) that describes appearance of an intense band in the CD spectrum of cholesteric liquid-crystalline dispersions may be rewritten as,

$$\Delta A_{\upsilon_i , \upsilon_i^*} \sim N \times f(C_{NA}), f^*(C_{CC}), f_j(\alpha), f_j^*(\alpha^*),$$

$$f_k (\overline{\varepsilon_{||} / \varepsilon_m}; \overline{\varepsilon_{\perp} / \varepsilon_m}), f_l (\upsilon_i), f_l^*(\upsilon_i^*), f(T) \times M \qquad (5)$$

where: $\Delta A_{\upsilon_i} = \Delta A_L - \Delta A_R$ is the circular dichroism at frequency υ_i; $\Delta A_{\upsilon_i}^*$ ($\Delta A_L - \Delta A_R$ is the circular dichroism at frequency υ_i^*; C_{CC} is the concentration of the colored biologically active compound bound to the DNA nitrogen bases; $\varepsilon_{||}, \varepsilon_{\perp}$ are the mean longitudinal and transverse components of dielectric permeability of the DNA molecule added with biologically active compound; α^* is the angle of inclination of the biologically active compound molecule with respect to the helical axis of DNA. Other parameters are defined above.

Analysis of expression (5) allows a few practically important conclusions to be drawn.

1. At least two intense bands are expected in the CD spectrum when the colored biologically active compound reacts with double-stranded DNA molecules forming liquid-crystalline dispersions. One of the bands will still be in the absorption region of the DNA nitrogen bases (υ_i) while the other will appear in the region (υ_i^*) where the chromophores of the colored compound absorb.

2. The sign of this new band will coincide with the sign of the band typical of DNA nitrogen bases if the compound is between nitrogen bases [the inclination angle (α^*) with respect to the DNA helix is $\sim 90°$]. If the compound is located on the DNA molecule so that its inclination angle is within 0-54°, the intense band in the CD spectrum has to have a sign, which is opposite to the sign of the band in the region of absorption of nitrogen bases.

3. The change of direction of the cholesteric twist of DNA liquid-crystalline dispersions can be accompanied by a simultaneous change of signs for both the bands in the CD spectrum.

4. Because the formation of the DNA liquid-crystalline dispersions takes place under fixed conditions, and angles of inclination of nitrogen bases and the biologically active compound are fixed, expression (5) can be written in a more simple form:

$$\Delta A_{\upsilon_i, \; \upsilon_i^*} \sim N \times f(C_{NA}), N_1 \times f^*(C_{CC}) \qquad (6)$$

The greater the DNA concentration used for the formation of liquid-crystalline dispersions, the higher the amplitude of the second band. At definite fixed DNA concentration:

$$\Delta A_{\upsilon_i^*} \sim N_1 \times f^*(C_{CC}) \qquad (7)$$

The greater the concentration of DNA-bounded compound, the higher the amplitude of a band in the CD spectrum in absorption region of this compound.

There are some limitations for expression (7). First, it is limited by the concentration (C_{CC}^{min}) of colored molecules. (It implies an existence of a minimal concentration of the molecules above for which a cholesteric ordering takes place). Second, it is limited by the (C_{CC}^{max}) concentration of the colored compound. (This indicates the existence of a maximal concentration of the colored compound that does not destroy DNA properties such as α or α^* values, mean dielectric properties, etc.). Above this concentration, according to expression (3), the probability for formation of a cholesteric structure begins to decrease. One can expect here diminishing of amplitudes for both bands in the CD spectrum specific to a liquid-crystalline dispersion formed from the DNA complexed with biologically active compounds. This observation shows that only under definite conditions, i.e. concentration of biologically active compounds, is the amplitude of the band in the CD spectra proportional to C_{CC}.

Therefore, appearance of intense bands in the CD spectrum specific to DNA liquid-crystalline dispersions with chemical compounds added allows not only the presence of compounds to be detected and their concentration in the medium under study estimated, but also establishes the mode of location for compounds on the DNA matrix.

Biosensing units based on the DNA liquid-crystalline dispersions operate as follows: DNA nitrogen bases "recognize" and "address" compounds, whereas the CD-band allows the results of these processes to be visualized. The described type of biosensing unit is suitable for "group determination" of compounds. Specificity

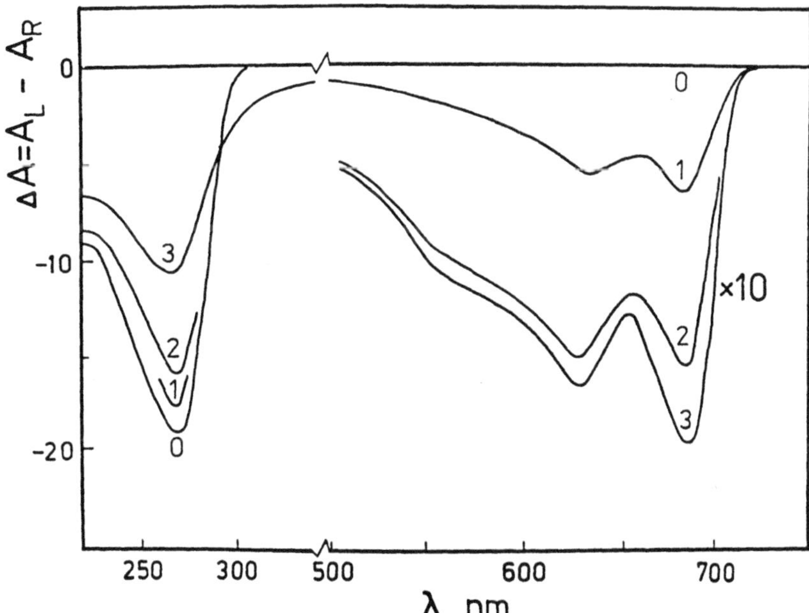

Figure 8. The CD spectra of liquid-crystalline dispersions formed from [DNA-MX] complexes. (0) $r = 0$, (1) $r = 0.037$, (2) $r = 0.08$, (3) $r = 0.11$. 0.3 M NaCl, $C_{PEG} = 150$ mg ml^{-1}, mol. mass of PEG = 4000. ΔA in cm, 1 cm = 1 x 10^{-4} optical units. The r value is the relation of the molar concentration of MX molecules bound to DNA nitrogen bases.

of determination depends on differences in binding constants and in optical properties of biologically active compounds. A few examples are given below.

Figure 8 compares the CD spectra of DNA liquid-crystalline dispersions upon adding antitumor compound—mitoxanthrone, i.e. 1,4-dihydroxy-5,8-bis[[2-[(2-hydroxyethyl)amino]-ethyl]amino]-9,10-anthracenedyone (MX) [24]. Adding of MX is accompanied by two sets of optical effects: (1) appearance of a new band in the CD spectrum, and (2) decrease of the band specific to the DNA liquid-crystalline dispersion although an intense negative band ($\lambda \sim 260$ nm) is present under all conditions. In accordance with the concept above, a new

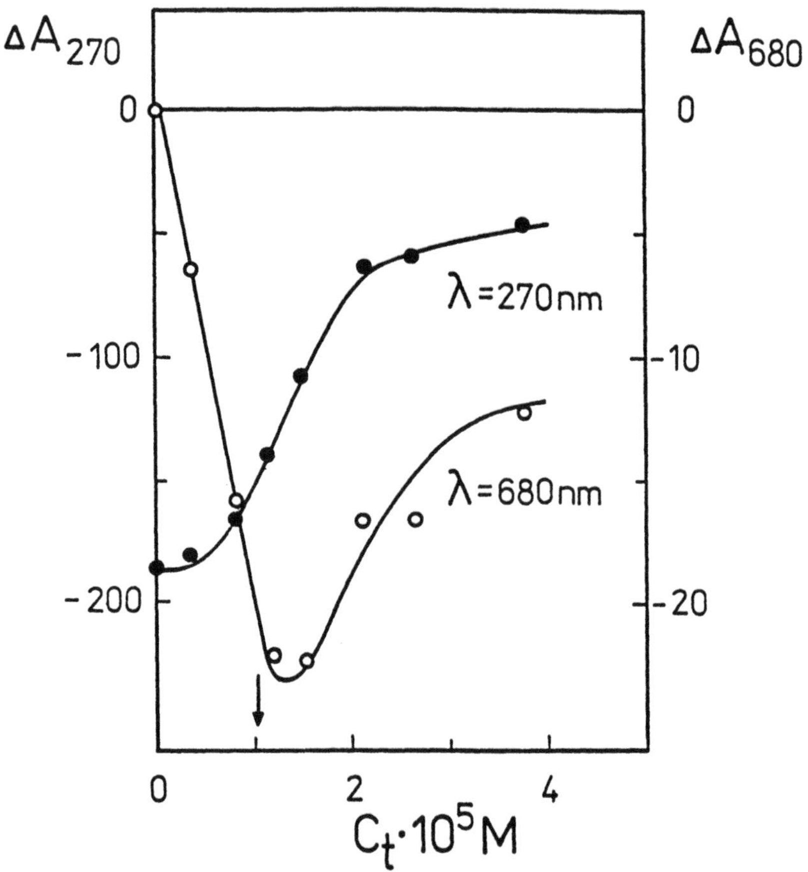

Figure 9.   Dependence of the amplitude of the bands in the CD spectra (λmax = 270 nm and λ_{max}= 680 nm) for liquid-crystalline dispersions formed from [DNA-MX] complexes versus concentration of MX. ΔA_{270} and ΔA_{680} in mm, 1 mm = 1 x 10^{-5} optical units. The C_t value is total concentration of MX added to solution. The arrow indicates the MX concentration limit detected by this method.

negative band appears in the region of absorption of MX ($\lambda_{max} \sim 680$ nm). The shape of this band resembles the shape of the absorption band. An appearance of the intense band in the CD spectrum in the absorption region of MX clearly shows that MX molecules are interacting with DNA. The negative sign of the band in the CD spectrum proves that MX molecules are located on DNA so that the angle of inclination (α^*) is $\sim 90°$; this is possible when molecules are located between nitrogen bases of DNA (intercalating between nitrogen bases). Figure 9 compares dependencies of amplitudes of negative bands (at $\lambda = 270$ nm and $\lambda = 680$ nm) versus total concentration of MX added to solution. These dependences are complex in their character. However, these dependencies show that there is a region of concentration where the amplitude of the negative band in the CD spectra is directly proportional to its concentration (up to the total concentration of $\sim 1 \times 10^{-5}$ M). Under these conditions the DNA structure is not distorted by the

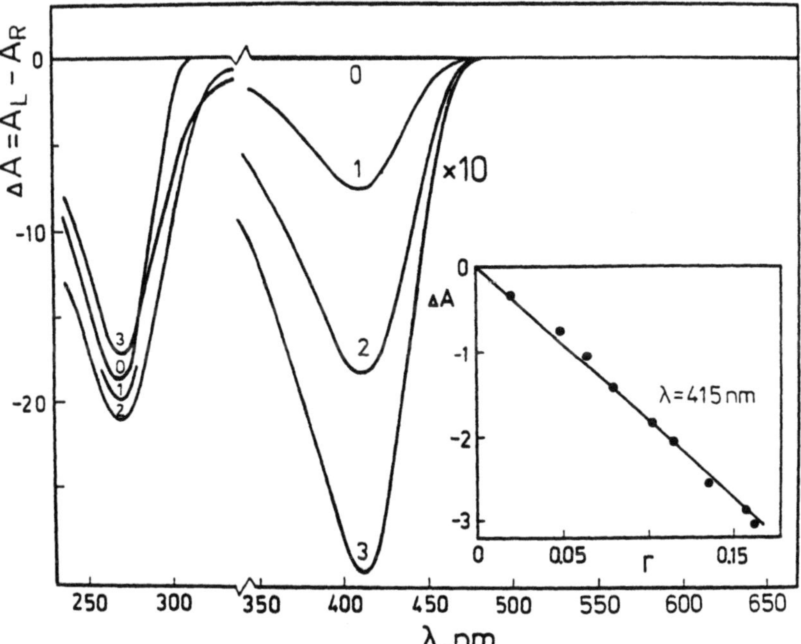

Figure 10. The CD spectra of liquid-crystalline dispersions formed from [DNA-BS] complexes. (0) $r = 0$, (1) $r = 0.049$, (2) $r = 0.102$, (3) $r = 0.163$. 0.3 M NaCl, $C_{PEG} = 150$ mg ml^{-1}, mol. mass of PEG = 4000. ΔA in cm, 1 cm = 1×10^{-4} optical units. The r value is the relation of the molar concentration of BS molecules bound to DNA to the molar concentration of DNA nitrogen bases. *Inset:* dependence of the amplitude of the band in the CD spectra ($\lambda_{max} = 415$ nm) of liquid-crystalline dispersions of the [DNA-BS] complexes versus r value.

interaction of MX with nitrogen bases. Hence, these dependencies allow conditions for MX determination to be chosen. The results obtained imply that the a biosensing unit based on DNA liquid-crystalline dispersions allows us to establish the presence of MX and MX concentration in the solution and also the mode of location of MX molecules on the DNA matrix.

Figure 10 illustrates the situation for the antitumor drug—9,10-anthracenedicarboxaldehyde-bis[(4,5-dihydro-1H-imidazole-2-yl)hydrazone]dihydrochlridbisanterene (BS) [25]. In contrast to MX, after adding BS to the DNA liquid-crystalline dispersion the amplitude of negative band in the CD spectra does not change significantly. The intense band in the CD spectrum in the absorption region of BS shows that this compound interacts with DNA. Besides, the negative sign of the band in the CD spectrum in the absorption region of BS proves that BS molecules are located between nitrogen bases of DNA (intercalating between nitrogen base pairs). Finally, the experimentally measured amplitude (ΔA) of the negative band in the CD spectrum in the absorption region of BS is directly proportional to the concentration of DNA-bound BS molecules. The correlation between r and C_t values allows us to establish the presence and BS concentration in the solution as well as the mode of location of BS on the DNA molecules.

Thus the biosensing unit based on linear, double-stranded DNA liquid-crystalline dispersions can be used for analytical purposes. The minimal concentration of colored compounds interacting with DNA molecules, which is being established now, is equal to ~ 10^{-6} - 10^{-7} M.

4. SUMMARY

The background for constructing biosensing units on the based on linear double-stranded DNA molecules is described above. A limitation that can influence optical properties of DNA liquid-crystalline dispersions is evident: it is the relatively low stability of these dispersions. From "model" consideration to real biosensors based on the principles considered, it is necessary to solve two problems:

1. To stabilize the properties of DNA liquid-crystalline dispersions. The preferable way consists of their immobilization in a swollen polymeric matrix. It can be added that although this problem is not considered theoretically, the same experimental data are in favor of its solution.
2. To construct a portable transducer of the optical signal(s) generated by biosensing units.

It can be expected that the solution of these problems would allow the design of biosensors for determination of biologically active compounds differing in structures as well as in mechanisms of the interaction with DNA molecules. Such biosensors could be used in laboratories and in industry.

REFERENCES

[1] Downs, M.E.A., Kobayashi, S., Karube, I. (1987). New DNA technology and the DNA biosensor. *Anal. Letters*, **20**, 1897-1927.

[2] Matthews, J.A., Kricka, L.J. (1988). Analytical strategies for the use of DNA probes. *Anal. Biochemistry*, **169**, 1-25.

[3] Yevdokimov, Yu.M., Skuridin, S.G. (1990). Biosensing units based on nucleic acids. (Russian ed.). In: *Itogi nauki i tekhniki* (Biotekhnologiya) (Yegorov, A.M., Ed.) Vol. 26, pp. 134-161. Moscow.

[4] Ngeh-Ngwainbi, J., Suleiman, A.A., Guilbault, G.G. (1990). Piezoelectric crystal biosensors. *Biosensors & Bioelectronics*, **5**, 13-26.

[5] Luong, J.H.T., Mulchandani, A., Guilbault, G.G. (1988). Developments and applications of biosensors. *Trends in Biotechnology*, **6**, 310-316.

[6] Fawcett, N.C., Evans, J.A., Chien, L.-C., Flowers, N. (1988). Nucleic acid hybridization by piezoelectric resonance. *Anal. Letters*, **21**, 1099-1114.

[7] Van Brunt, J., Klausner, A. (1987). Pushing probes to market. *Biotechnology*, **5**, 211-221.

[8] Gebhart, F. (1989). Molecular devices introduces its biosensor-based DNA assay. *Gen. Eng. News*, **9**, 1.

[9] Yevdokimov, Yu.M., Skuridin, S.G., Chernuha, B.A. (1992). Biosensing units based on double-stranded nucleic acid liquid-crystalline dispersion. (Russian ed.). *Biotekhnologiya*, **5**, 103-109.

[10] Yevdokimov, Yu.M., Skuridin, S.G., Salyanov, V.I., Rybin, W.K. (1991). General principles of creating biosensing units based on double-stranded nucleic acid liquid crystals. In Lazarev, P.I. (ed.), *Molecular Electronics*, (Lazarev, P.I., Ed.), pp. 317-329. Kluwer Academic Publishers, Netherlands.

[11] Yevdokimov, Yu.M., Skuridin, S.G., Salyanov, V.I. (1988). The liquid-crystalline phases of double-stranded nucleic acids *in vitro* and *in vivo*. *Liquid Crystals*, **3**, 1443-1459.

[12] Saeva, F. (1979) Cholesteric liquid crystal- induced circular dichroism. In: *Liquid crystals. The forth state of matter,* (Saeva, F., ed.), pp. 249-273. Marcel Dekker, New York.

[13] Sackman, E. Voss, J. (1972). Circular dichroism of helically arranged molecules. *Chem. Phys. Letters*, **14**, 528-532.

[14] Mason, S.F., Peacock, R.D. (1973). The location of the ^1L_b transitions of anthracene by dichroism techniques. *Chem. Phys. Letters*, **21**, 406-408.

[15] Yevdokimov, Yu.M., Skuridin, S.G. (1988). Encapsulated nucleic acid liquid crystals (Russian ed.) *Dokl. Akad. Nauk S.S.S.R.*, **303**, 232-235.

[16] Spada, G.P., Brigidi, P., Gottarelli G. (1988). The determination of the handedness of cholesteric superhelices formed by DNA fragments. *J. Chem. Soc.*, **14**, 953-954.

[17] Osipov, M.A. (1985). Molecular theory of solvent effect on cholesteric ordering of lyotropic polypeptide liquid crystals. *Chem. Phys.*, **96**, 259-269.

[18] Sonin, A.S., Belyakov, W.A. (1982). The Optics of Cholesteric Liquid Crystals, (Russian ed.), p. 360. Nauka, Moscow.

[19] Adamczyk, A. (1989). Phase transition in freely suspended smectic droplets. Cotton-Mouton technique, architecture of droplets and formation of nematoids. *Mol. Cryst. Liq. Cryst.*, **170**, 53-69.

[20] Yevdokimov, Yu .M., Skuridin, S.G., Akimenko, N.M. (1984). The liquid-crystalline microphases of low molecular mass double-stranded nucleic acids and synthetic polynucleotides. (Russian ed.) . *Vysokomolek. soyedineniya*, **26**, 2403-2410.

[21] Yevdokimov, Yu. M., Skuridin, S.G. Lortkipanidze, G.B. (1992). Liquid crystalline dispersions of nucleus acids. *Liquid Crystals*, **12**, 1-16.

[22] Yakovlev, D.Yu., Skuridin, S.G., Khomutov, A.R., Yevdokimov, Yu. M., Khomutov, R.M. (1992). Photochemical modification of nitrogen bases and spatial organization of the DNA lyotropic liquid crystals (Russian ed.). *Dokl. Akad. Nauk S.S.S.R.*, **325**, 393-396.

[23] Yevdokimov, Yu. M., Skuridin, S.G., Salyanov, V.I., Damaschun, G., Damaschun, H., Misselwitz, R., Kleinwachter, V. (1990). Effect of platinum(II) chemotherapeutic agents on properties of DNA liquid crystals. *Biophys. Chem.*, **35**, 143-153.

[24] Kapuscinski, J., Darzynkiewicz, Z., Traganos, F., Myron, R. (1981). Interactions of a new antitumor agent, 1,4-dihydroxy-5,8-bis[[2-[(2-hydroxyethyl)amino]-ethyl]amino]-9,0-anthracenedione, with nucleic acids. *Biochem. Pharmacol.*, **30**, 231-240.

[25] Citarella, R.V., Wallace, K.C., Murdock, K.C., Angier, R.B., Durr, F.E. and Forbes, M. (1982) Activity of novel anthracenylbishydrazone,9,10-anthracenedicarboxaldehyde-bis [(4,5-dihydro-1-H-imidazol-2-yl)hydrazone] dihydrochloride, against experimental tumor in mice. *Cancer Res.*, **42**, 440-444.

LANGMUIR–BLODGETT MONOLAYERS AS A BASIS FOR ADVANCED OPTICAL BIOSENSORS

Alexander P. Savitsky and Valery V. Savransky

OUTLINE

Advances in Biosensors
Volume 3, pages 165-190.
Copyright © 1995 by JAI Press Inc.
All rights of reproduction in any form reserved.
ISBN:1-55938-535-9

ABSTRACT

Photophysical principles and labels for the detection of biospecific interactions at surfaces and the application of Langmuir-Blodgett films of antibodies as sensitive elements of biosensors based on the total internal reflection of excited fluorescence and on the plasmon resonance principle are described. The immunological activity of IgG molecules and steric hindrances for antigen binding in the Langmuir-Blodgett films are discassed.

1.   PHOTOPHYSICAL PRINCIPLES AND LABELS FOR THE DETECTION OF BIOSPECIFIC INTERACTIONS AT SURFACES

Antibody-antigen reactions offer the possibility of unique specificity in recognition of a wide spectrum of studied compounds. High sensitivity in the detection of these compounds may be achieved if an effective system is used for the detection reaction. The most sensitive and convenient immunochemical methods use an antibody-antigen reaction at a solid surface. Development of the theory and practice of this approach for the last 20 years is extensively reviewed in the literature [1-3]. For immunosensors, the solid-surface mediated detection of the antibody-antigen complex is the most reliable and simple approach for the development and manufacture of remote fiber-optical devices.

Two detection principles are the most promising for optical immunosensors. The first is surface plasmon resonance (SPR), which permits unlabeled antibodies and antigens to be used. It simplifies the assay procedure and preparation of reagents and allows the development of a reversible sensor operating on-line. The sensitivity of this method, however, is limited by nonspecific adsorption of sample proteins due to the unrecognizability of specific and nonspecific adsorption. This is a common problem for all immunological methods in which unlabeled compounds are used. Second, if labeled compounds are used, it is quite easy to recognize specific and nonspecific adsorption of the antibodies or antigens. In this case the method based on the effect of total internal reflection of excited fluorescence (TIRF) is the simplest. While the sensitivity of this type of

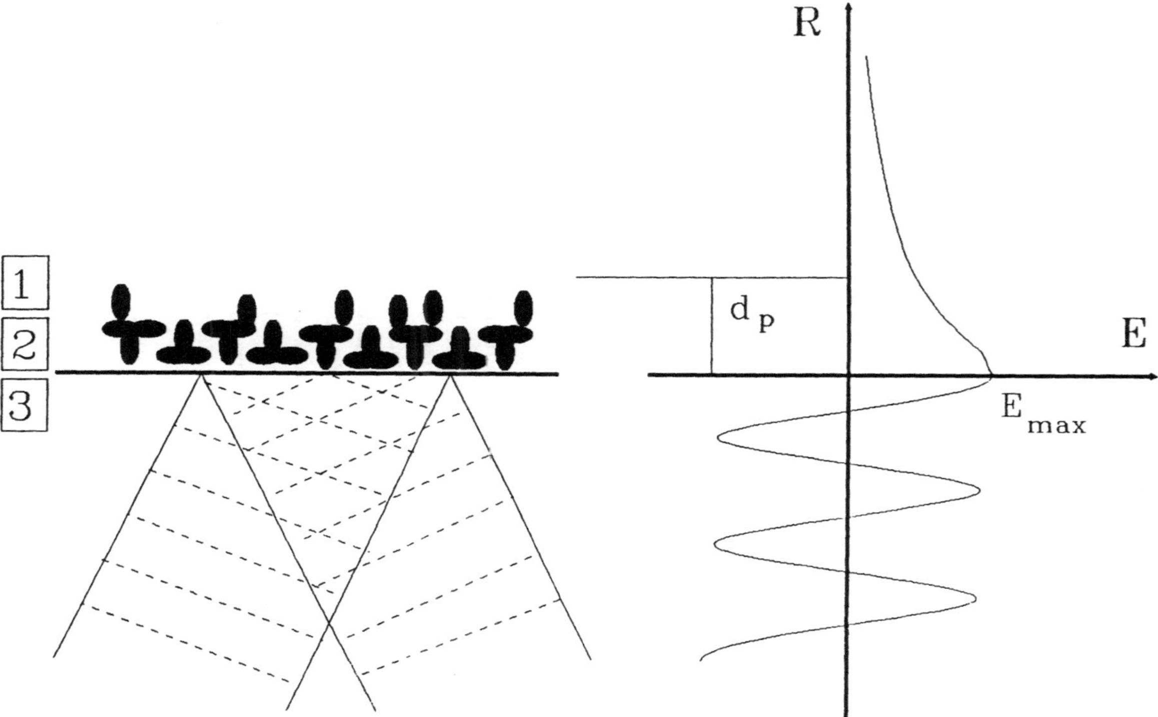

Figure 1. Schematic illustration of the TIRF or SPR immunosensors. R is the distance in the test solution, d_p is the evanescent field penetration depth and E is the electric field amplitude. (**1**) Labeled (unlabelled) antibodies (antigens). (**2**) Immobilized antibodies (antigens). (**3**) Waveguide.

immunosensor will be higher than SPR sensors, but it is difficult to produce a reversible rather than a single-use device.

The investigation of the structure and the functional activity of Langmuir-Blodgett (LB) films of immunoglobulins are of great interest for the development of new biosensors. The use of IgG monolayers as immunosensor sensing elements based on TIRF and SPR is effective for two reasons. First, the Langmuir technique allows the formation of two-dimensional condensed monomolecular antibody films and their deposit onto solid supports. The sensitivity of immunosensors, based on optical detection principles [4,5], can only be high when fluorescence-labeled antibodies (antigens) are evenly distributed along their working surface and within the area of the evanescent surface wave propagation (Figure 1). Second, this technique makes it possible to form oriented monolayers of antibodies whose immunological activity exceeds that of antibodies immobilized by traditional methods [6].

In this review we describe the principle of SPR methods, new labels for potential use in TIRF sensors and the methods used in the preparation of protein LB-films and their immunological properties.

2. SURFACE POLARITON WAVES IN LANGMUIR-BLODGETT FILMS

There is a method of analyzing metallic surfaces based on optical excitation of surface plasmon waves (SPW). This method, using surface plasmon resonance effects (SPR), turned out to be far more sensitive to the properties of the surface and interfaces than others relying upon the effect of light reflection. The method (SPW) has been available since Fano accounted for Wood's anomalies in diffraction gratings [7], but was put into wide practice only after the works of Otto [8] and Kretschmann [9] who demonstrated the possibility of exciting SPW on a smooth surface. These authors have developed a scheme known as the method of frustrated total internal reflection (FTIR) (Figure 2).

In Otto's scheme the light passes through three media with dielectric constants ε_1, ε_0, and ε_2. The medium with dielectric constant ε_1 is a triangular prism. If the incidence angle at the interface of media with dielectric constant ε_1 and ε_2 is larger than arcs in $(\varepsilon_1, \varepsilon_2)^{1/2}$ (the angle of total internal reflection), then the tangential component of the wave vector k_x is maintained when passing through the interface, whereas its normal component k_{0n} becomes an imaginary value. That corresponds to the process of the evanescent electromagnetic wave in a medium with e_0, on its moving in the direction of Z. Since $k > (\omega/c)\varepsilon_0^{1/2}$, resonant excitation of surface plasmons is possible at the interface of media with dielectric constants ε_1 and ε_0. As a result, the intensity of the light reflected from the interface will be reduced (frustrated total internal reflection). The thickness of the gap between media with ε_1 and ε_2 should not be large because of a fast attenuation

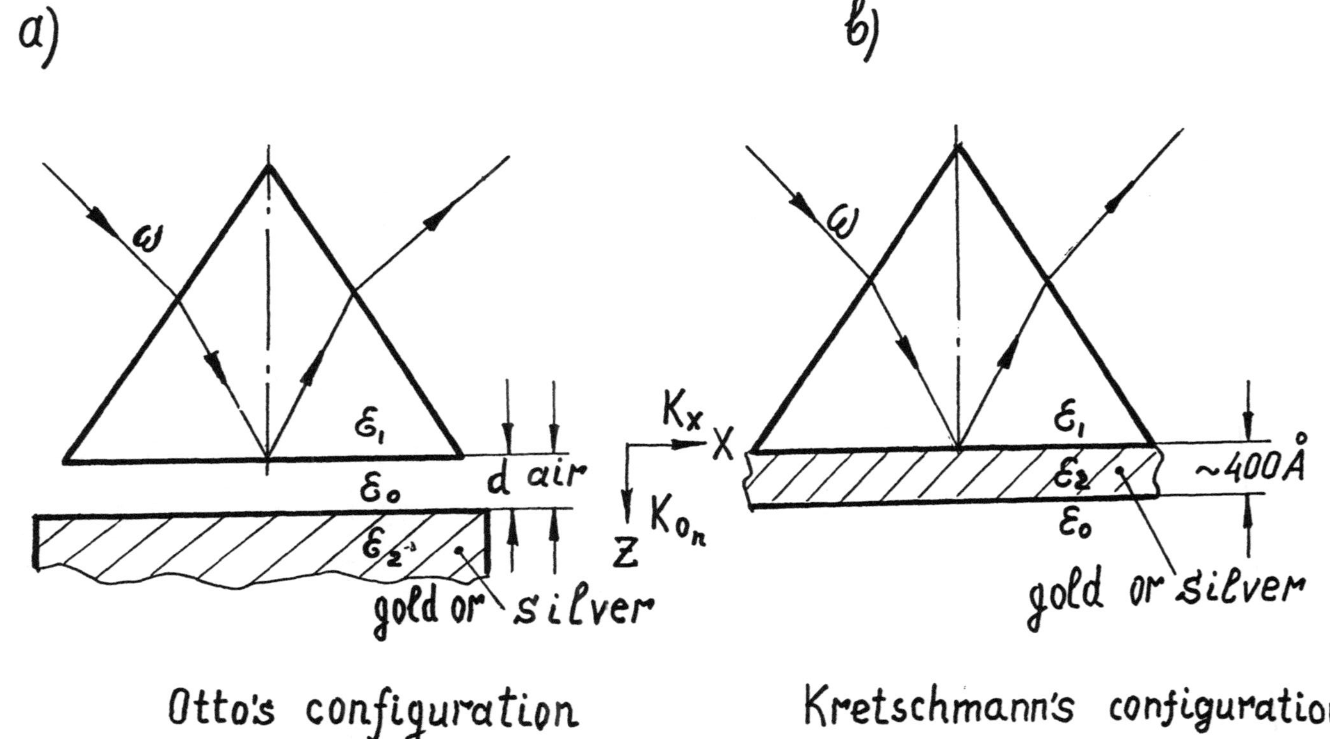

Figure 2. Optical scheme of observation of the effect of surface plasmon resonance: (a) Otto scheme, (b) Kretschmann scheme.

of the electromagnetic wave, since it is necessary for the wave to have a large enough amplitude when reaching the medium with ε_2. On the other hand, the attenuation should be noticeable in order to excite SPW at the interface of media with ε_0 and ε_2. Optimal conditions are met when the gap thickness is comparable with the incident radiation wavelength. In Kretschmann's scheme, to excite the surface wave at the interface of media with ε_2 and ε_0 and to attain resonance, a prism is necessary with dielectric constant ε_1 and the gap thickness is chosen proceeding from the same considerations used in Otto's scheme.

The SPW excited by the FTIR method, can be observed in two ways. The first is based on detecting the intensity of the light reflected from the interface of media with ε_1 and ε_0 (or ε_2 and ε_0), depending on the incidence angle. On exciting SPW, the intensity of the reflected light will decrease, i.e. the p-polarized wave reflection index will change. In the second approach, the roughness at the interface of media with ε_2 and ε_0 and the inhomogeneous dielectric constant ε_2 in the intermediate layer, give rise to SPW scattering and, respectively, to light scattering emission decay. The SPW can be recorded by observing the dependence of the reflected light intensity versus the incidence angle of the p-polarized wave.

In the above approaches, the development of SPW has a resonant character (SPR) and strongly depends on the SPW dispersion properties, which, in turn, depend on the properties of the material and on a thin layer adjacent to the surface.

The presence of an intermediate layer with dielectric constant $\varepsilon_2 > 1$ will give rise to changes in the dispersion curve, and in case of a fixed external light frequency, the resonance will be attained at another value of the wave vector of surface plasmons. The vector in case of resonance will be equal to $\mathbf{S} = \mathbf{S}_0 + \Delta\mathbf{S}$, where $\mathbf{S}_0$ is an unperturbed value of the wave vector of plasmons at which the resonance is attained, and $\Delta\mathbf{S}$ is a contribute of the intermediate layer. These additions to frequency and to the wave vector are proportional to the thickness of the intermediate layer. In FTIR experiments, the real part is responsible for the angular shift of the SPR curve, and the imaginery one for the SPR curve broadening [10].

The amplitude of the resonance curve is usually given in a normal form, R_p / R_p where R_p is the reflection index for the p-polarization, and R_s is the reflection index s-polarization, the latter actually independent of the light incidence angle of the prism base. The measurements performed by the FTIR approach can give an idea of the intermediate layer absorption band if resonance curves are taken at various frequencies of the incident light.

We have carried out these FTIR experiments by the Kretchmann approach, using a glass (n=1.5163) prism with a 48 nm-thick gold film deposited on its base. The experimental approach is as follows. The light from a He-Ne laser with an intensity 6 mW at 633 nm reaches the gold film through the prism. The intensity of the reflected light was recorded by the photodetector. The total internal reflection angle was 41.8°. The wave vector met the SPR curve at an angle of 43.12° ± 0.05°. The SPR curve half-width was 0.67° ± 0.1°.

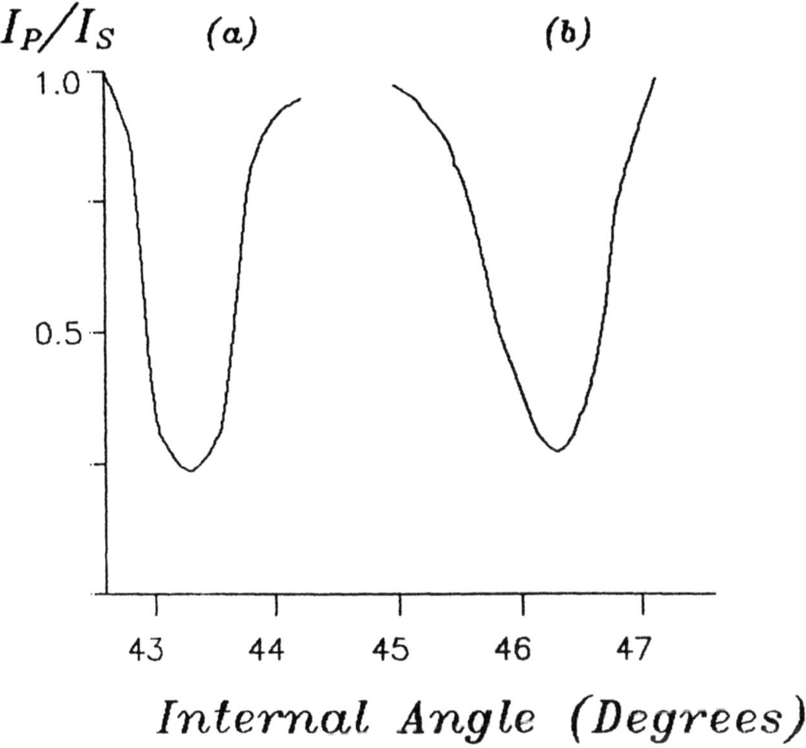

Figure 3. SPR data for gold film (**a**) and coated of bacteriorhodopsin LB-film (**b**).

After measuring the parameters of the gold film, half of the surface was covered with a monolayer of bacteriorhodopsin (bR) molecules by the Langmuir-Blodgett method; then the SPR parameters were again measured. The open part of the gold film featured the resonance as before at an angle of 43.12°. The half-width of the resonance curve of the gold film with deposited bR was 0.78° ± 0.1°. The difference in angles at which the two resonances were observed, was 2.9° ± 0.1°. The dependences obtained are presented in Figure 3[11-13].

3. NEW LABELS FOR POTENTIAL USE IN TIRF

3.1 Background Photoluminescence of Biological Samples and Fiberoptics

As discussed before, the main problem in background signals for SPR is the nonspecific adsorption of sample proteins due to the unrecognizability of specific and nonspecific adsorption. This problem can be solved only by immunochemical and chemical methods used in the preparation of a sensing surface.

But in the case of TIRF, certain problems in background photoluminescence can be solved by the choice of fluorescent label. Biological specimens are characterized by high light-scattering and background fluorescence, which cause a decrease in the sensitivity of assays. In the case of fiber-optical biosensors, an additional problem of fiber luminescence and transmittance for excitation and emission wavelength arises. These unfavorable effects can be minimized by selecting a label with high absolute sensitivity, appropriate wavelength for excitation, emission and resolution between them, and a time-resolved mode of measurement.

3.2 Wavelength-Resolved Mode of Measurement

All biological samples have a very strong background luminescence in the UV-range. In a visible range, the background signal depends upon excitation and emission wavelength [14]. The use of porphyrins as fluorescent labels has a series of advantages over traditional fluorescence labels. The porphyrins can be excited in the range of 380-600 nm, and emission at 550-650 nm permits the effective minimization of light-scattering and background fluorescence of samples [15]. When the standard series of fluorimeters is used, porphyrin labels permit a 10- to a 100- fold increase in sensitivity, by comparison with traditional labels like FITC, TRITC, DANS, and porphyrin labels suitable for assays of thyroxine [16], digitoxine [17] and influenza virus A [18]. The far-visible and near-infrared spectral regions (600-1000 nm) are areas of low interference, where only a few classes of molecules exhibit significant absorption or fluorescence [19]. This spectral region is of special interest to invasive biosensors. Aluminium-sulfophthalocyanin has outstanding fluorescence and physicochemical properties [20], and is one of the most promising for use in biosensors.

3.3 Time-Resolved Mode of Measurements

The time-resolved mode of measurement, or phosphorymetry, is one of the most sensitive techniques (competing with the radioactive tracer by its sensitivity parameters) and was introduced to immunoassay by Soini and co-workers [14]. The method reduces background signals to a minimum and measures weak luminescence of a labeled compound in complex biological samples. Phosphorymetry principally differs from other methods of fluorescent assay at the stage of luminescence measurement. After the sample is excited by a short pulse of light (1-2 μs), measurement is delayed for hundreds of (Figure 4) microseconds. During this delay the background signal almost completely decays and a signal of long-lived label is measured devoid of noise. The registration cycle may be repeated many times (1000 times and more) to increase the integral signal. The scattered light effectively coincides with the time profile of the flash, but post-luminescence of fibers, optical elements of the sensors, also contributes to the background signal. The decay time of these signals doesn't exceed 100-200 μs.

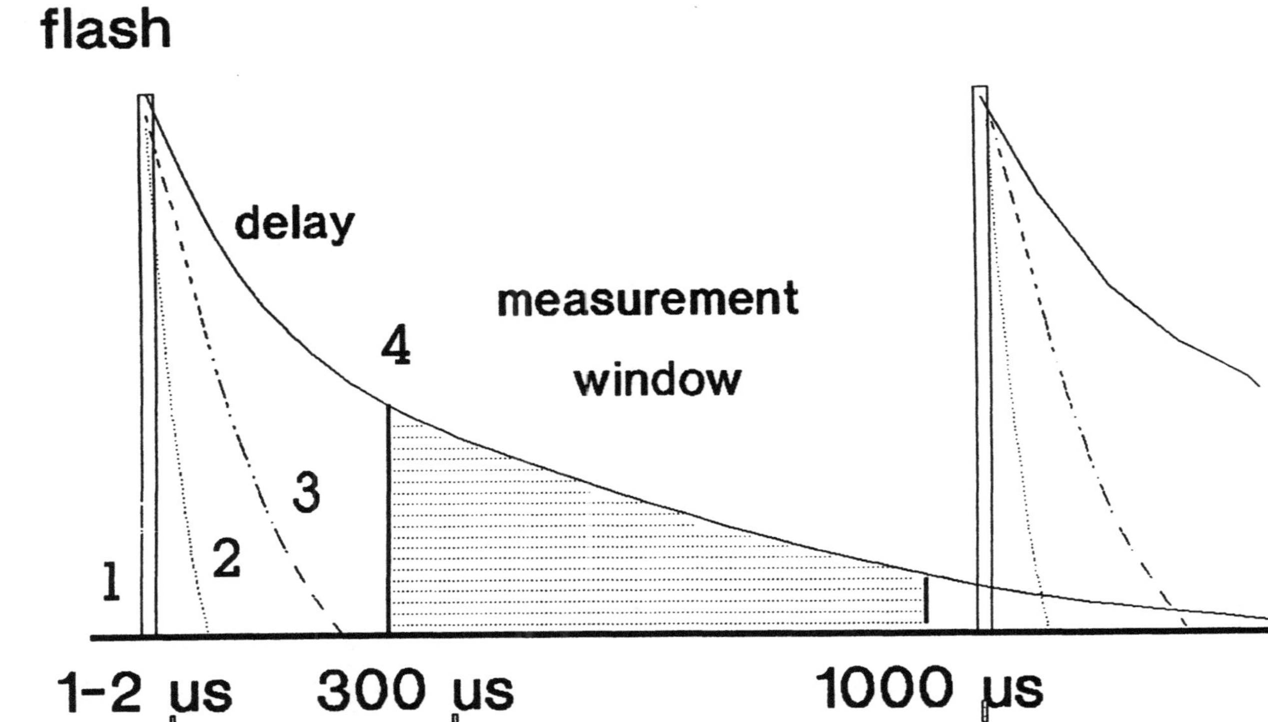

Figure 4. The principle the time-resolved mode of measurements. (1) flash; (2) background luminescence of samples; (3) background luminescence of plastic; (4) luminescence of label.

Therefore, despite several requirements for luminescent labels of special character, the introduction of a microsecond delay line before registration of a label reduces the background signal to a minimum. Such an approach enables an increase in sensitivity of detection by 2-3 orders, as compared with stationary fluorescent spectroscopy, if an effective long-lived luminescent label is used.

3.4   Measurement of Metalloporphyrin Room Temperature Phosphorescence in Solution

The use of metalloporphyrin phosphorescence at room temperature as a label in biosensors permits the regime of time discrimination of background signals to be applied. Pallachium coproporphyrin (Pd-CP) has excellent phosphorescence characteristics. In detergent solution it displays high quantum yield (O.17) and molar extinction under excitation at 395 nm (200,000 M^{-1} cm^{-1}) [21]. This results in detection limits as low as 10^{-13} M for the label (Figure 5) in micellar solution. Metalloporphyrines (Zn, Al, etc.) exhibiting delayed fluorescence in aqueous solutions at room temperature have also been suggested to be used as labels [22]; however, the quantum yields of those labels are rather low.

It is possible to use plastic fibers in combination with porphyrin labels due to the high transmition of plastic fibers at the wavelength of excitation. Excitation in the visible range of 520-580 nm is suitable for the use of diode lasers as a light source.

If fiber-optical immunosensors are to be used as miniature remote sensing support for traditional solid-phase immunoassays, it is necessary to optimize conditions for metalloporphyrin room temperature phosphorescence measurement.

Long-term luminescence of organic compounds in solutions is affected by quenchers of triplet states; dissolved oxygen plays the main role in this process. However, in aqueous solutions this problem is solved rather simply. Dissolved oxygen can be removed from aqueous solutions by the addition of sulfite or bisulfite [23]. Due to a chemical reaction with sulfite, dissolved oxygen can be rapidly and effectively removed for a long time, which is sufficient to perform phosphorescent measurement. Depending upon the concentration of sulfite, the intensity of phosphorescence in an open tube goes unchanged for 3 or 4 hours.

Nonionic detergents have a significant effect on the phosphorescence quantum yield of metalloporphyrins in water solutions. As demonstrated in our previous work, the porphyrin macrocycle is localized in the micellar hydrophobic nucleus [24] and thus protected against quenching by water. Optimal concentration for Triton X-100 is between 1 and 2% (Figure 6). Under this condition, the quantum yield is 10 times higher than for a water solution at the same pH. It is interesting to note that in contrast to porphyrin fluorescence [24] there is no simple correlation in phosphorescence quantum yield to a critical concentration of micellation for Triton X-100 (0.02% v/v). The phosphorescence quantum yield significantly increased after critical concentration of micellation for Triton X-100 was reached.

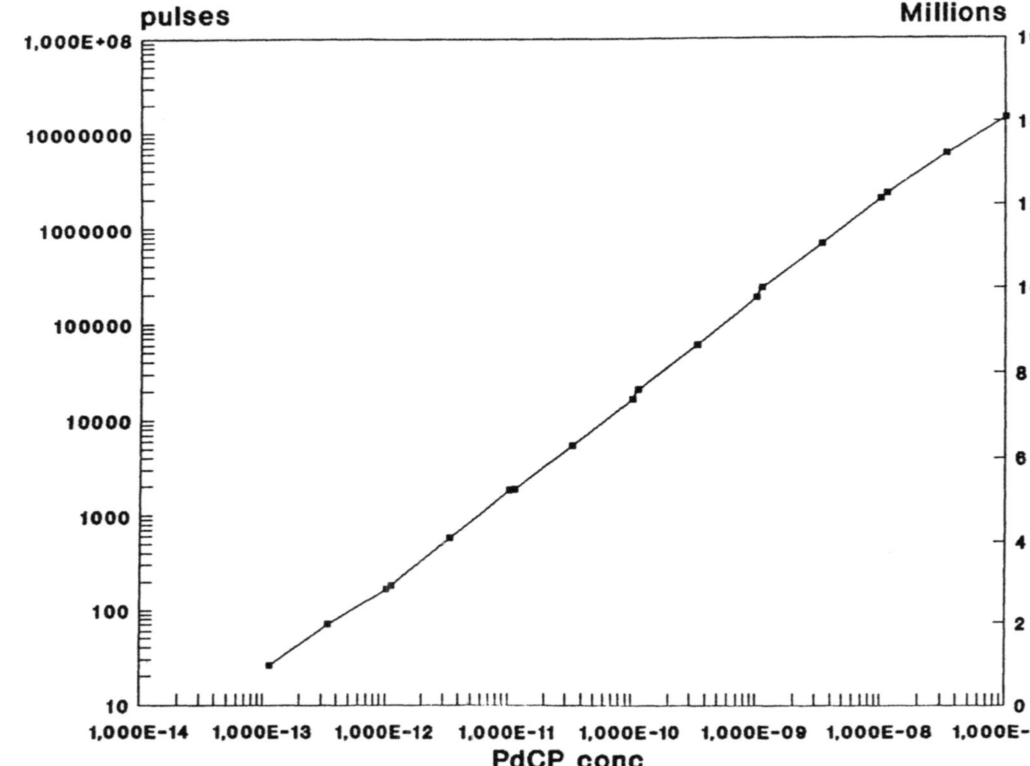

Figure 5. Detection limit and dynamic range for Pd-coproporphyrin at pH 7.4 in 1% Triton X-100, 10 mM phosphate buffer, 10 mg/ml Na_2SO_3.

175

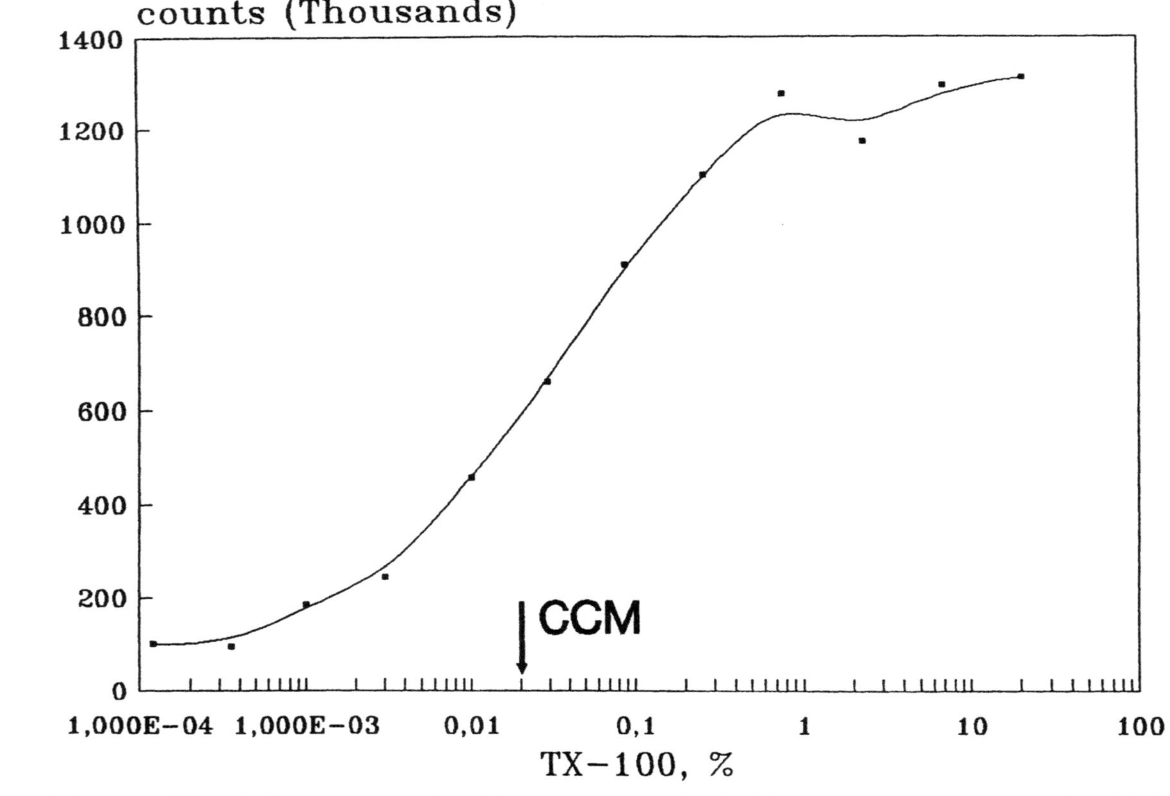

Figure 6. Influence of Triton X-100 on the intensity of phosphorescence of 1 nM Pd-coproporphyrin solution in 10 mM phosphate buffer, pH 7.4, 10 mg/ml Na_2SO_3.

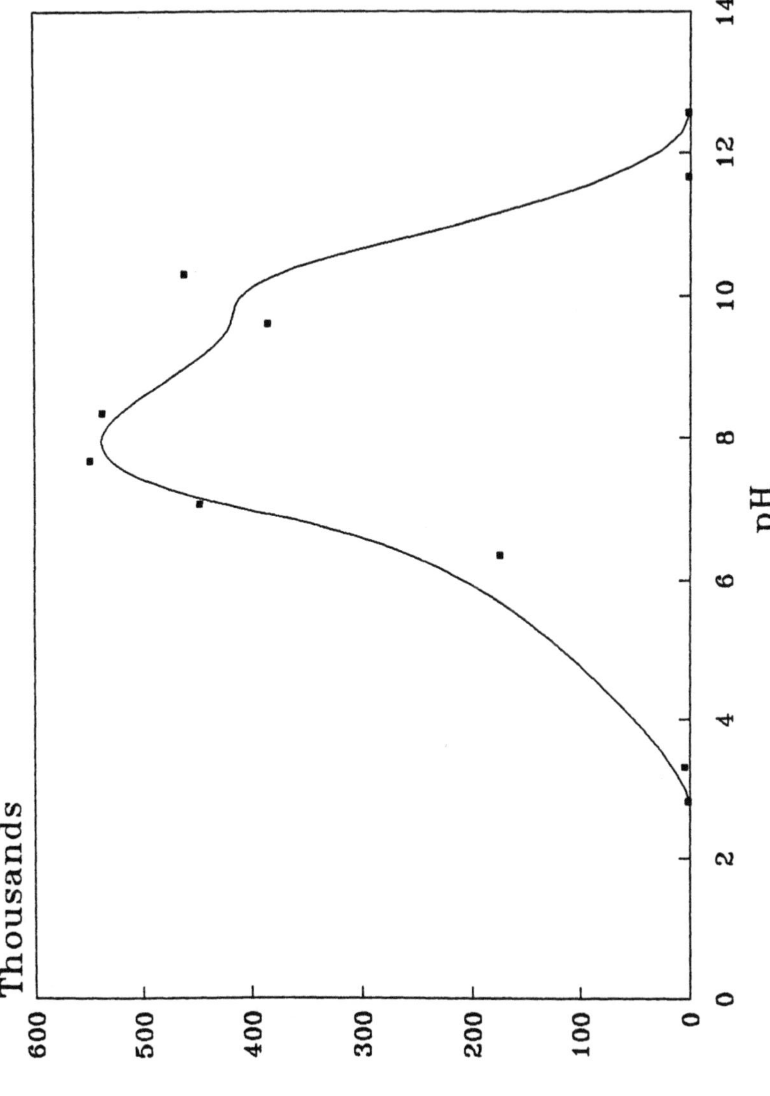

Figure 7. pH dependence of Pd-coproporphyrin phosphorescent intensity in 1% Triton X-100, 10 mM phosphate buffer, 10 mg/ml Na_2SO_3.

But it was also possible to use a very low concentration of detergent (0.02-0.06%), which did not destroy antibody-antigen interaction at the surface.

The pH value has a much greater effect on the phosphorescence quantum yield than detergent concentration (Figure 7). Phosphorescence intensity at acid pH is strongly decreased in comparison with neutral pH. The decrease in phosphorescent intensity for Pd- and Pt-coproporphyrins is similar to that in fluorescence intensity for free coproporphyrin at acid pH [25]. One possible explanation for this phenomenon is the protonation of coproporphyrin carboxylic groups, which was demonstrated with several carboxylic porphyrins and their esters in micellar solution [26]. The nature of the decrease in the intensity of phosphorescence at alkali pH is not clear, but it is clear that at neutral pH it is possible to measure the phosphorescent signal without any precautions.

3.5   Solid-Surface Measurement of Metalloporphyrin Phosphorescence

A recent discovery of solid-surface room temperature phosphorescence of metalloporphyrins [27] offers a new possibility for the development of single-use phosphorescence sensors, in which a disposable fiber with a definite amount of specifically bound metalloporphyrin-labeled compound can be stored and repeatedly measured for a long time (up to one year).

Solid-surface phosphorescence analysis is very sensitive and selective for organic trace analysis, but moisture and oxygen quenching can be a problem in room temperature phosphorescence. Several solid materials, including filter paper, silica gel, aluminium oxide, silicone rubber, sodium acetate, potassium bromide, and cellulose are used in solid-surface phosphorescence. Phosphorescence signals are strongly affected by such factors as solution chemistry, evaporation of the solvent, drying of the solid surface, and phosphor-solid matrix interactions. Sample compartments for solid-surface phosphorescent measurements are blown with dry and sometimes hot nitrogen or air [28].

In contrast to the conditions of measurement described above, the intensity of the phosphorescent signal of Pd-CP conjugated with streptavidin at the dry surface after specific interaction in the presence of oxygen, corresponds to 248% of the signal of the same conjugate in a deoxygenated micellar solution. The chemical nature of the solid surface (polystyrene, glass, silica, surface of cells) has no effect on the intensity of the phosphorescent signal in the presence of oxygen. The nature of coupling of Pd-CD to proteins (covalent or specific interaction with antibodies), as well as the nature of attachment of proteins to the surface (adsorption on the polystyrene or chemical fixation of Langmuir-Blodgett films of Pd-CP conjugate on the silica) have no effect on the phosphorescent intensity.

Different polymeric substances, like mounting medium for microscopy, can be used for the enhancement of the phosphorescent signal from the surface. After polymerisation of the mounting medium, the phosphorescent signal of Pd-CP is even higher than for the dry surface (277%) [29]. Moreover, coating of the fiber by polymer after immunoreaction protects the sensing and signaling surface against mechanical damage and prolongs the storage time for use of the signaling disposable unit of the sensor.

4. LANGMUIR-BLODGETT FILMS OF PROTEINS

The Langmuir method allows for superordered films with a definite thickness and dense molecular packing.

LB-technology provides:

1. Films homogeneous in thickness within 10-1000 Å on the support surface of a required area.
2. Successive deposition of monolayers allow for the creation of structures with pre-given characteristics; the choice of composition and deposition technology and the optimization of physical and chemical properties in monolayers, are made according to a definite programme.
3. The use of mixed layers—monomer, polymer, and combined structures (for instance, skeletization technology) allows films to be obtained with a pre-given distribution and developing structures with a specific surface topography.

Thus, almost any analytical function or variation of functions can be realized in LB-films.

4.1 New Approach for Preparation of Protein LB-Films: Reverse Micelles

There is a problem in making LB-films with active hydrophilic proteins because they are denatured during the process of their spreading on the surface of a water-air interface, or moving from the surface to the volume.

The method of formation of highly ordered LB-films from hydrophilic substances has been proposed with cytochrome C, and rhodamin 6G as the model, with no preliminary modification of their structure [30]. This method is based on solubilization of the substance by surfactant reversed microemulsions in organic solvents. This is followed by spreading of the microemulsion on the surface of the water. The structure of water bubbles with the included protein molecule is shown in Figure 8. We used the well known substances—surfactant Aerosol OT (AOT), cytochrome C and rhodamin 6G for studying the main principles of the film construction. The solubilization of protein prevents of its denaturation at the instant of spreading and by the organic solvents.

Protein microemulsion systems are commonly used for various proteins and low molecular weight substances [31,32]; their structures are very variable and easily controlled under varying external conditions. The destruction of water

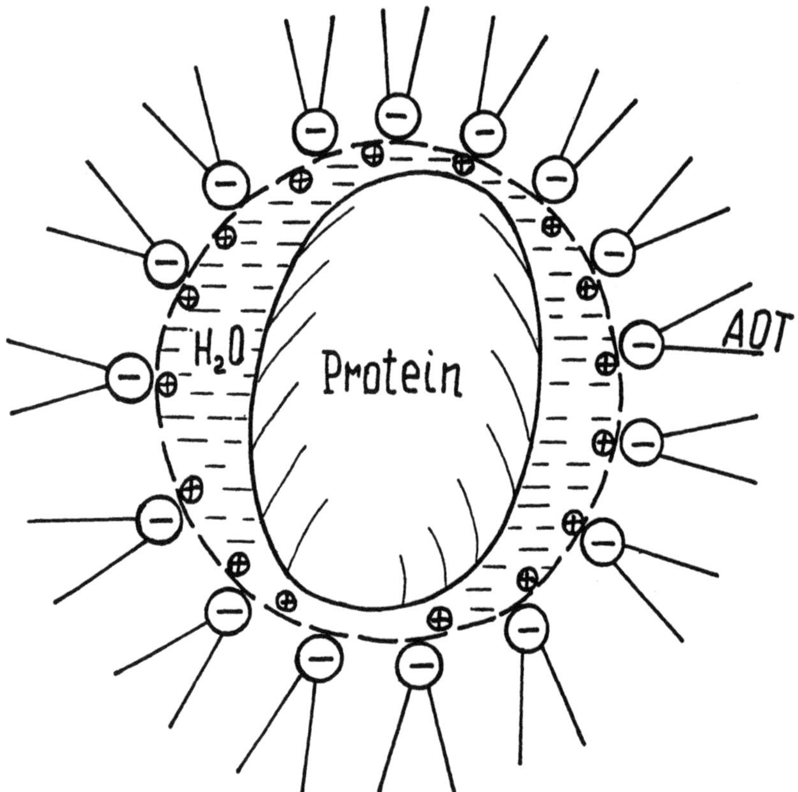

Figure 8. The structure of the water microemulsion bubble with a solubilizated protein molecule.

bubbles in a microemulsion is followed by evaporation of the organic solvent on the water surface after spreading. Without any solubilized substances, the surfactant molecules of the microemulsion are dissolved in the water phase, but using cytochrome or rhodamin, the structure appears similar to that in Figure 9.

The ordered structure (Figure 9) can be more successfully made by using a periodical moving of the barrier from the initial position to one with a surface tension of up to 30 mN/m. This procedure speeds up the destruction of microemulsion bubbles and helps to attain more effectively the state of equilibrium.

As we established, the essential electrostatic attraction has to be between solubilized and surfactant molecules. The neutralization of molecular charges through interaction among molecules makes substances more hydrophobic, whereas their excess is dissolved in the water phase.

Langmuir films were transferred onto a solid support in X- and Z-transfers. Using the method of small-angle X-ray scattering (see Figure 10), it was estab-

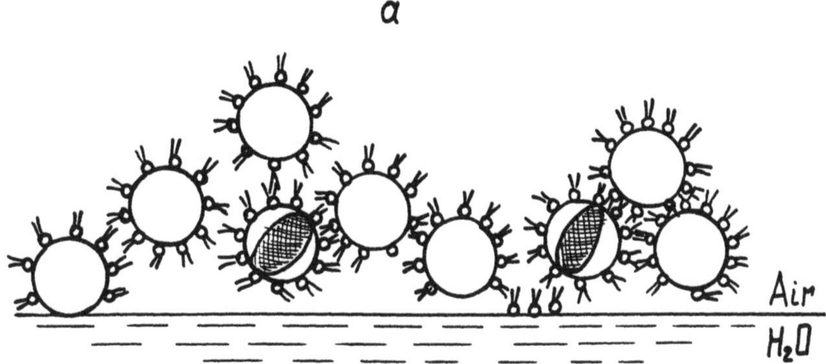

Figure 9. The film structure on the surface of water: (**a**) at the instant of microemulsion spreading; (**b**) after attaining equilibrium state.

lished that the protein films are highly ordered, and lattice space was found to be 55±1 Å. The structure of multilayer film is shown in Figure 11.

The microemulsion-spreading method is useful for forming LB-films from non-traditional polymers and surfactants with small-volume polar groups.

4.2 Preparation of LB-Films of Antibodies

For a successful application of the Langmuir technique in immunosensors, it is necessary to determine antibody LB-film parameters at the liquid-gas phase interface at various surface pressures, and also to determine the molecular orientation of antibodies in the monolayer. The immunological activity of the antibody LB-films at solid surface was determined with the help of RIA [6], ELISA [33], and gravimetrical methods [34,35], and was shown to depend on the surface pressure

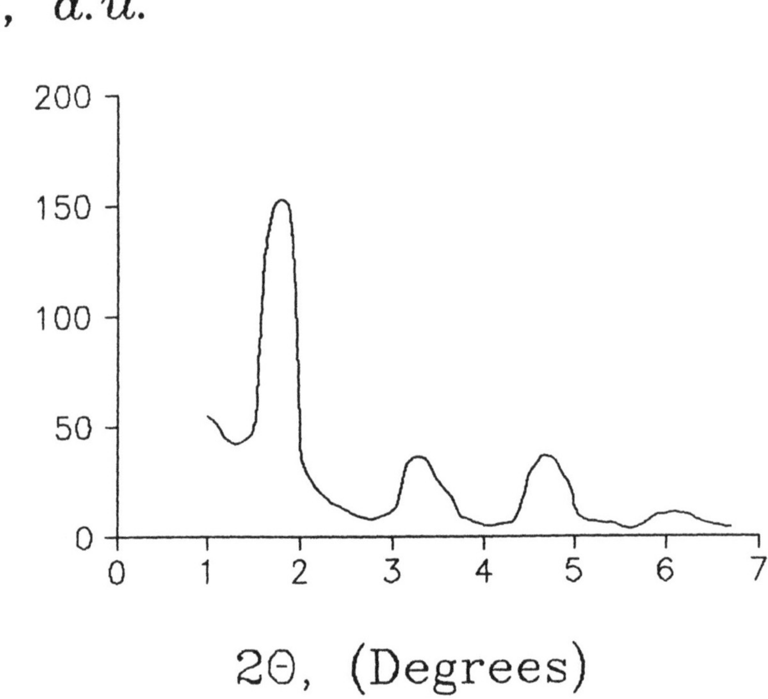

Figure 10.   X-rayogram of the protein multilayer film on a solid support.

in the immobilized layer, on the number of deposited layers, and on the immobilization method. At the same time, quantitative estimations of the IgG monolayer structure and density made by different authors are contradictory. Evaluated by the ellipsometry method, the monolayer thickness is between 50 Å [33] and 108 Å [6]. The methods applied in the investigation of protein LB-films, such as small-angle X-ray scattering or light absorption at 280 nm, can be used mainly for multilayer structures. However, no sufficient order in protein multilayer has been observed, as diffractional images do not contain more than three reflexes [34].

Our approach to the study of LB film structure consists of using fluorescence-labeled antibodies and antigens for the quantitative evaluation of immunological properties of IgG LB-films [36,37]. As an antigen, Pd-CP was used, which is intensely phosphorescent at room temperature [21]. The immunological activity of the Pd–CP-specific IgG monolayer was determined on the basis of its binding with Pd–CP, conjugated to proteins of different molecular weight. The use of dual-labeling methods—fluorescent europium labelling for antibodies and phosphorescent metalloporphyrin labeling for antigens—made it possible to determine antibody concentration in the monolayer and their functional activity simultaneously within the same experiment. Quantitative measurements were

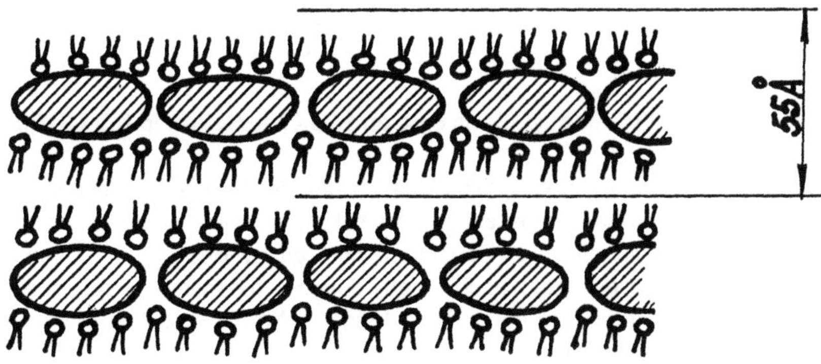

Figure 11. The structure of a multilayer film obtained by using the data from small angle X-ray scattering.

facilitated by the fact that during the measurement procedure both labels (europium and porphyrin) dissociated from the support surface into the solution.

Monoclonal antibodies to Pd-CP were obtained at the Institute of Biochemistry in Moscow [38]. Mice were immunized by conjugate of Pd-CP with bovine serum albumin. After fusing mouse immune lymphocytes with mielome cells Sp2/O, cell growth of the hybridomes was observed in 30 cases, which positively responded to the ELISA test. [Pd-CP conjugated with soybean inhibitor of trypsin (SIT) was used as an antigen in adsorption on the microplates]. Eighteen hybridomes belonged to IgM class of immunoglobulines, and five to the different subclasses of IgG. Affinity constants were determined for four monoclonal antibodies, which demonstrated an effective interaction with free Pd-CP. This was done by using solid-phase immunoassay in which the concentration of non-bound antigen was estimated. The results calculated by a Scatchard plot are presented in

Table 1.   Dissociation Constants of Monoclonal Antibodies to Pd-Coproporphyrin

Name of antibodies	Class of Ig	Dissociation Constant of Pd-CP	Dissociation Constant of Pd-CP/SIT
D3.F5	IgG 1	4.8×10^{-10}M	3.2×10^{-10}M
D5.E3	IgG 1	5.0×10^{-9}M	5.5×10^{-9}M
D9.E6	IgG 2B	2.1×10^{-8}M	1.5×10^{-8}M
C2.5B	IgG 1	—	7.3×10^{-9}M

Table 1. As seen from the table, dissociation constants for the interaction of monoclonal antibodies with Pd-CP/SIT and with free Pd-CP are similar. Monoclonal antibodies D3.F5 (3D) were used to study the LB-film structure.

4.3   The Area Per One IgG Molecule and Its Orientation in the LB-Film

To calculate the area per one IgG molecule in the film, we made direct measurements of the concentration of the europium-labeled antibodies. Using the Schaefer method, the antibody film was transferred onto quartz disks of fixed size, at varying surface pressure, and was covalently bound to the surface as described in [37]. The concentration of europium ions, dissociated from the labeled antibodies at the surface, were measured in the fixed volume by the DELFIA method. Knowing the degree of europium labeling of antibodies (1/1), it is possible to get the exact value of the area per IgG molecule in the immobilized film.

Taking into account the X-ray data of antibody molecule structure [39,40], one can propose two extreme boundary positions of IgG molecules in the monomolecular film. These may be characterized by the area per molecule in the film equal to 7000 Å (the smallest one) and 12,600 Å (the biggest one), respectively (see Figure 12). The first figure corresponds to the perpendicular position of the IgG molecule and the second to its planar position to the film plane.

Figure 13 shows how the area per molecule in the IgG monolayer depends on the value of the surface pressure. Analyzing this dependence, one can come to the conclusion that at a pressure under 25 mN/m the estimated dimensions of the area per IgG molecule in the monolayer exceeds its real dimensions with any orientation of the molecules. When the surface pressure reaches 30 mN/m, the area per molecule in the monolayer is 12,900 Å which could be associated with the larger cross-section of the molecule. At 40 mN/m the area per molecule is 7900 Å, which could be associated with the smaller cross section of the IgG molecule.

4.4   Immunological Properties of Monoclonal Antibody LB-Films

Several orientations as related to the film plane of antibody molecule Fab fragments responsible for the antigene binding, are possible. They can be oriented predominantly in the water-subphase direction, or in the gas-phase direction, or in both directions.

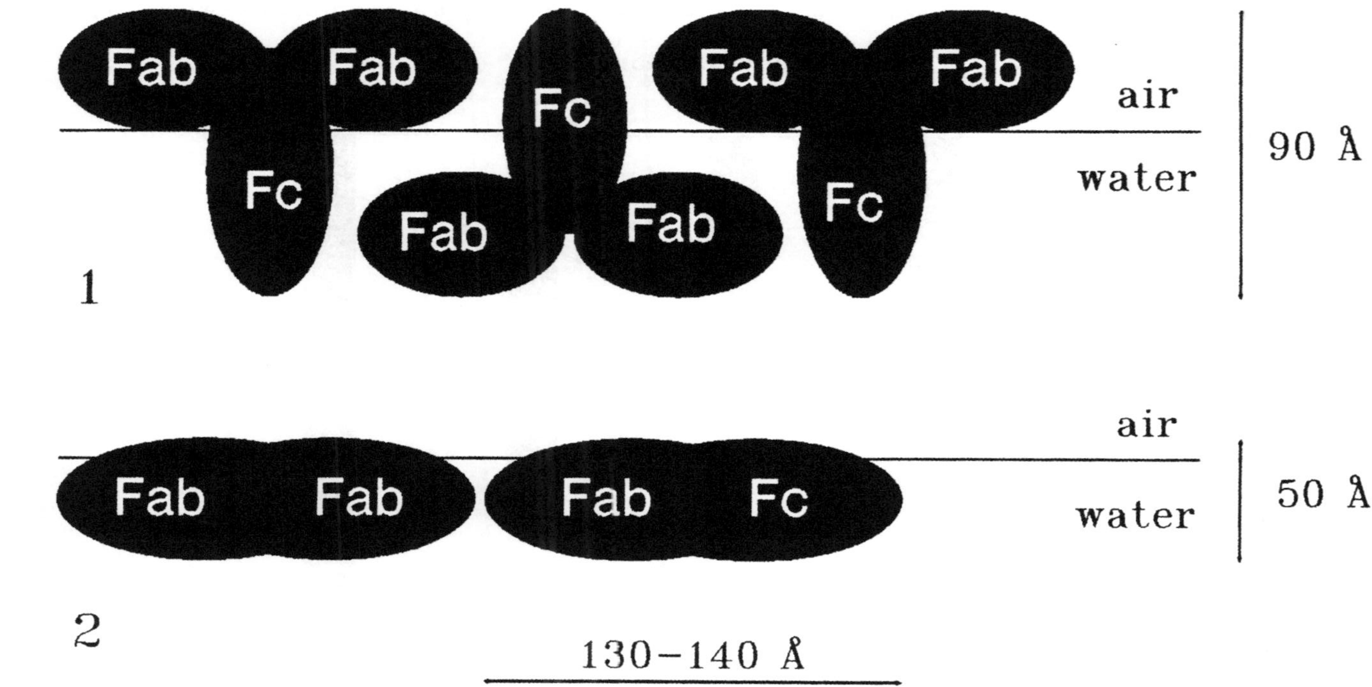

Figure 12. Models of different IgG molecules orientation in the monolayer at the water-air interface. (**1**) The larger cross-section is perpendicular to the interface. (**2**) The larger cross section is parallel to the interface.

185

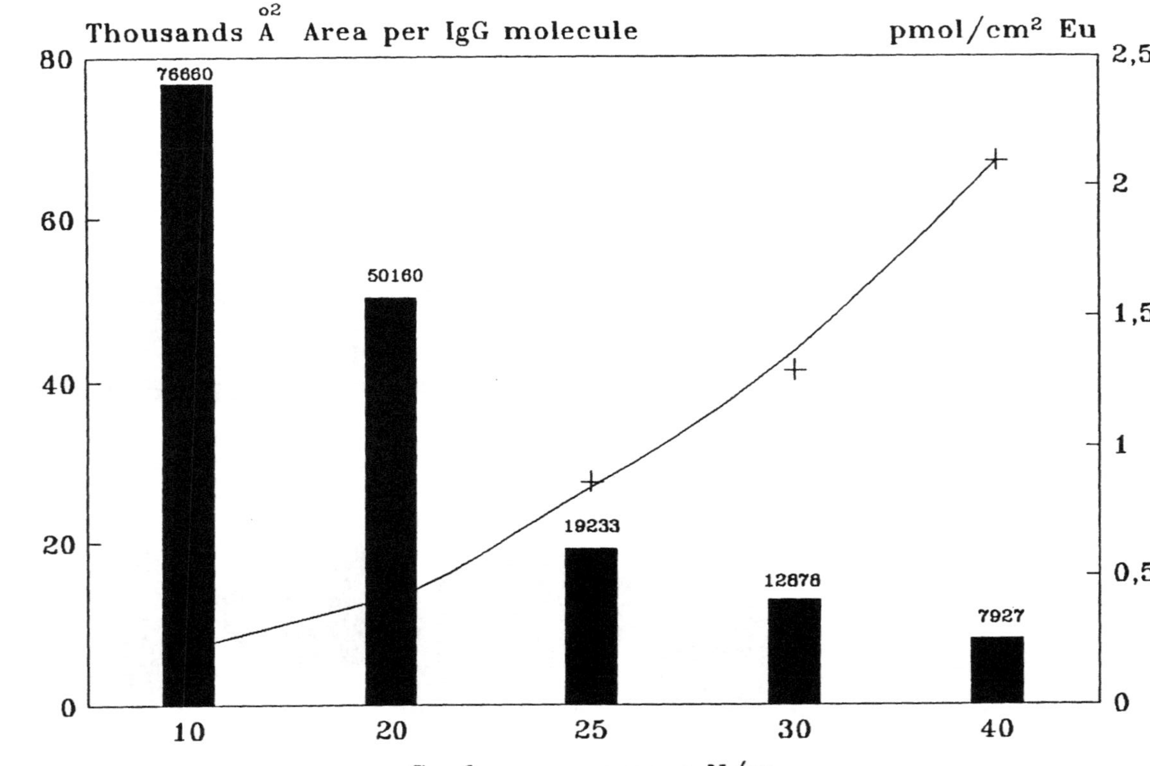

Figure 13. The dependence of the size of the area per one IgG molecule in the immobilized LB film on the surface pressure in the formed monolayer.

Table 2. Specific Binding of Different Molecular Weight Antigens
by LB-Films of Monoclonal Anti-PdCP Antibodies

Antigens	Molecular Weight of the Antigen	Degree of Conjugate PdCP Labelling	Immunological Activity of the Antibody Active Sites in the LB Film (%)
PdCP	761	—	46.0
PdCP-SIT	25000	1.63	8.27
PdCP-TG	669000	77.0	0.25

The immunological activity of the LB-film of monoclonal mouse IgG was evaluated on the basis of its binding free PdCP and PdCP conjugated with different proteins: thyroglobulin, MW 669,000 (TG); and soybean inhibitor of trypsin, MW 25,000 (SIT). The area of the antibody LB-film deposited onto the quartz disk on both sides was 0.392 cm. The amount of the IgG immobilized onto the disk surface was determined from fluorescence measurements and was 0.8 pmol per disk.

The maximal values of the amount of antigen specifically bound by LB-film are given in Table 2. When binding a free antigen, 46% of the binding centers (IgG have two binding centers) of the LB-film antibodies are active, which corresponds to 0.74 pmol of PdCP per disk. This can be explained by the fact that the antibodies are oriented in two opposite directions in the LB-monolayer and only antibodies directed to the solution can bind the antigen, in contrast to antibodies directed to the support. With the increase of the molecular weight of the antigen, the value of the specific binding decreases to 0.133 and 0.004 pmol per disk for PdCP-SIT and PdCP-TG conjugates, respectively. This means that 8.27 and 0.25% of the binding centers of antibodies in the monolayer are active. That could be accounted for by steric hindrances associated with binding. On the basis of the data obtained, the conclusion can be made that in binding a low-molecular antigen, LB-film antibodies preserve half of their immunological activity. This could probably be explained by the fact that the antibodies were oriented in two opposite directions in closely packed films, and one of these orientations was not favorable for the inter-action with antigens (see Figure 12). However, the amount of the antigen thus bound depends on its molecular weight and seems to be conditioned by the steric hindrances occurring during high-molecular antigen interaction with LB-film.

The investigations carried out show that antibody LB-films can be effectively used as sensing elements for immunosensors.

ACKNOWLEDGMENTS

This research was carried out in the laboratory of molecular immunology of the Institute of Biochemistry and in the laboratory of laser biophysics of the General Physics Institute of the Russian Academy of Sciences. The authors of this paper would like to thank T.

Dubrovsky, M. Demcheva and K. Mantrova from the laboratory of molecular immunology, L. Belovolova, S. Valyansky, T. Konforkina, from the laboratory of laser biophysics; and Prof.G.V.Ponomarev from the Institute of Biophysics of the Ministry of Health of Russia for their participation and assistance in this research.

REFERENCES

[1] Complementary Immunoassay. Collins, W.P. (Ed.) (1988). John Wiley & Sons, New York.

[2] Ngo, T.T. and Lenhoff, H.M. (Eds.) (1985). *Enzyme-Mediated Immunoassay.* Plenum Press, New York.

[3] Hemmila, I.A. (1991). *Applications of Fluorescence in Immunoassays.* John Wiley & sons, New York.

[4] Andrade, J.D., Vanwagenen, R.A., Gregonis, D.E., Newby, K., Lin, J.N. (1985). Remote fibre optic biosensors based on evanescent-excited fluoroimmunoassay: concept and progress. *IEEE Trans.*, **32**, 1175-1179.

[5] Attridge, J.W., Daniels, P.B., Deacon, J.K., Robinson, G.A., Davidson, G.P. (1991). Sensitivity enhancement of optical immunosensors by the use of a surface plasmon resonance fluoroimmunoassay. *Biosens.Bioelectron.*, **6**, 201-214.

[6] Turko, I.V., Yurkevich, I.S., Chashchin, V.L. (1991). Langmuir-Blodgett films of immunoglobulin G for immunosensors. *Thin Solid Films*, **205**, 113-116.

[7] Fano, U. (1941). The theory of anomalous diffraction gratings and quasi-stationary waves on metallic surface. *J.Opt. Soc. Amer.*, **31**, 213-222.

[8] Otto, A. (1968). Exitation of nonradiactive surface plasma waves in silver by the method of Frustrated total reflection. *Ztschr.Physik.*, **216**, 313-324.

[9] Kretschmann, E. (1971). Die Bestimmung optisher Konstanten von Metallen durch Anregung von Oberflachenplasmaschwingungen. *Ztschr.Physik.*, **241**, 313-3244.

[10] Agranovich, V.M., Mills, D.L. (1985). *Surface polaritons*, pp. 185-189. Nauka, Moscow (Russian).

[11] Alekseev, A.S., Valynskiy, S.I., Prokhorov, A.M., Savransky, V.V. (1989). Giant Raman scattering of bacteriorhodopsin monolayers on surface plasmons. *Brief Communications on Physics*, **9**, 53-55 (Russian).

[12] Alekseev, A.S., Valynsky, S.I., Prokhorov, A.M., Savransky, V.V. (1990). Biological sensors bases on LB monolayers with registration by optical methods. *Proceedings of the USSR Academy of Sciences, ser.Phys.*, **54**, 1945-1947 (Russian).

[13] Valynsky, S.I., Konforkina, T.V., Savransky, V.V. (1992). Investigation of LB-films by angle- and frequency-angle absorbsion spectra of surface plasmon polaritons. Brief Communications on Physics, **5**, 3-6 (Russian).

[14] Soini, E., Hemmila, I. (1979). Fluoroimmunoassay: present status and key problems. *Clin.Chem.*, **25**, 353-361.

[15] Savitsky, A.P., Papkovsky, D.B., Berezin, I.V. (1987). Fluorescence immunoassay. Porphyrins as a new labels for immunoassay. *Dokl.Acad.Nauk. SSSR*, **293**, 744-748 (Russian).

[16] Papkovsky, D.B., Savitsky, A.P., Bykhovsky, V.Ya. (1988). Fluoroimmunoassay of thyroxine in human serum. *Probl.Endokrin.*, **34**, 58-61 (Russian).

[17] Krylova, S.M., Stafeeva, O.A., Savitsky A.P., Dikov, M.M., et.al. (1990). Solid-phase fluorescence immunoassay of digitoxine. *Voprosy Med.Khimii*, **36**, 47-49 (Russian).

[18] Ivanova, V.T., Stafeeva, O.A., Savitsky, A.P., Yakhno, M.A. (1988). Solid-phase fluorescence immunoassay of influenza virus A. *Voprosy Virusolog.*, **1988**, 362-365 (Russian).

[19] Patonay, G., Antoine, M.D. (1991). Near-infrared fluorogenic labels: new approach to an old problem. *Analyt.Chem.*, **63**, 321A-327A.

[20] Savitsky, A.P., Lopatin, K.V., Golubeva, N.A., Poroshina, M.Yu., et.al. (1992). pH dependence of fluorescence and absorbance spectra of free sulphonated aluminium phthalocyanine and its conjugate with monoclonal antibodies. *J. Photo-chem. Photobiol. B:Biol.*, **13**, 327-333.

[21] Savitsky, A.P., Papkovsky, D.B., Ponomarev G.V., Berezin, I.V. (1989). Phosphorescence immunoassay. Metalloporphyrins—alternative labels for rare earth fluorescent metal chelates? *Dokl.Acad.Nauk.SSSR*, **304**, 1005-1008 (Russian).

[22] Osin, N.S., Lepeschkina, N.V., Papkovsky, D.B., Rumyantseva, V.D., et.al. (1988). Russian Patent 1561042.

[23] Sidki, A.M., Smith, D.S., London, J. (1986). Direct homogeneous phosphoroimmunoassay for carbamazepine in serum. *Clin.Chem.*, **32**, 53-56.

[24] Savitsky, A.P., Vorobyova, E.V., Berezin I.V., Ugarova, N.N. (1981). Acid-base properties of protoporphyrin IX and its dimethyl ester and heme solubilized on surfactant micelles; spectrophotometric and fluorometric titration. *J.Coll.Interface Sci.*, **84**, 175-181.

[25] Papkovsky, D.B., Savitsky, A.P. Ponomarev, G.V. (1989). The fluorescence of porphyrins in detergent solution. *Zh.Prikl.Spektr.*, **51**, 786-790 (Russian).

[26] Savitsky, A.P., Ugarova, N.N., Berezin, I.V. (1979) Physicochemical properties of protoporphyrin IX and its dimethyl ester solubilized in micellar detergent solution. *Bioorg.Khem.*, **5**, 259-267 (Russian).

[27] Savitsky, A.P., Hemmila, I., Mantrova, K., Demcheva, M., et.al. (1993). Solid-surface measurement of room temperature phosphorescence of Pd-coproporphyrin and its application for time-resolved microscopy. *International Symposium on Biomedical Optics*, Budapest, September 1-5, 1993.

[28] Hurtubise, R.J. (1989). Solid-surface luminescence spectrometry. *Anal.Chem.*, **61**, 889A-895A.

[29] Mantrova, E.Yu., Demcheva, M.V., Savitsky, A.P., Hemmila, I. (1993). Solid-surface measurement of Pd-coproporphyrin labeled compounds phosphorescence. *International Symposium on Biomedical Optics*, Budapest, September , 1-5, 1993.

[30] Belovolova, L.V., Erokhin, V.V., Savransky, V.V. (1992). Inclusion of protein into Langmuir films from "water-in-oil" microemulsion. *Biological Membranes*, **9**, 765-770 (Russian).

[31] Martinek, K., Levashov, A.V., Khmelnitzky, Yu.L. (1981). Molecular mechanisms of protein solubilisation into organic solvent by surfactant. *Proceedings of the USSR Academy of Sciences*, **258**, 1488-1492 (Russian).

[32] Genkin, M.V., Davidov R.M., Belovolova, L.V. (1989). Catalytic and redox reactions of proteins and low molecular weight reactants in reverse micelles. *J. Mol. Catalysis*, **56**, 249-259.

[33] Ahluwalia, A., De Rossi, D., Monici, M., Schirone, A. (1991). Thermodynamic study of Langmuir antibody films for application to immunosensors. *Biosens. Bioelectron.*, **6**, 133-141.

[34] Erokhin, V.V., Kayushina, R.L., Lvov, Yu.M., Feigin, L.A. (1989). Protein Langmuir-Blodgett films as sensing elements. *Studia Biophys.*, **132**, 97-104.

[35] Dubrovsky, T.B., Erokhin, V.V., Kayushina, R.L. (1992). Gravimetric biosensors based on Langmuir-Blodgett films of immunoglobulin. *Biolog. Membr.*, **6**, 130-137 (Russian).

[36] Savitsky, A.P., Dubrovsky, T.B., Demcheva, M.V., Mantrova, E.Yu., et.al. (1993). Fluorescent and phosphorescent study of LB antibody films for application to optical immunosensors. In *Biomedical Optics '93*, (Katzir, A., Ed. *Proceedings of SPIE*, Bellingham, Washington (in press).

[37] Dubrovsky, T.B., Demcheva, M.V., Savitsky, A.P., Mantrova, E.Yu, et.al. (1993). Fluorescent and phosphorescent study of Langmuir-Blodgett antibody films for application to immunosensors. Biosensors (in press).

[38] Demcheva, M.V., Cherednikova, T.V., Savitsky, A.P., Mantrova, E.Yu., et.al (1991). Monoclonal antibodies to Pd-coproporphyrin. Russian patent N 1659477.

[39] Marquart, M., Deisenhofer, J. (1982). The three-dimensional structure of antibodies. *Immunology Today*, **3**, 160-167.
[40] Amit, A.G., Mariuzza, R.A., Phillips, S.E., Poljak, R.J. (1986). Three-dimensional structure of an antigen-antibody complex at 2.8 Å resolution. *Science*, **233**, 747-753.

LIGHT BIOSENSORS BASED ON BACTERIORHODOPSIN AND PHOTOSYNTHETIC REACTION CENTERS

A. A. Kononenko and E. P. Lukashev

OUTLINE

Advances in Biosensors
Volume 3, pages 191-211.
Copyright © 1995 by JAI Press Inc.
All rights of reproduction in any form reserved.
ISBN:1-55938-535-9

ABSTRACT

Retinal and chlorophyll proteins among the native light energy transducing complexes have definite advantages for use in biotechnology and molecular electronic applications as non-traditional active biomaterials. The unique combination of photochromic, photoelectric, and electrooptical properties of these proteins is of great fundamental interest and allows the development of new effective systems for high-speed processing of optical information, dynamic pattern recognition, and addressable data memory devices. Mono-and multilayer thin films (just dried or in polymer matrixes) of high-ordered bacteriorhodopsin (BR)- and bacterial reaction center (RC)-containing samples can be fabricated by Langmuir–Blodgett and electro-sedimentation methods on solid state optically transparent, current-conducting supports. General characteristics are described of the systems already realized, such as: disk optical information carriers based on BR operating in "record-read-erase" mode; electric field-controlled photochromic optical devices—thin films of BR between transparent electrodes; light-electricity transformers—spectrally selective and position sensitive charge-coupled optical detectors on BR and RC. The possible use of light-sensitive pigment proteins in biosensors is discussed.

1. INTRODUCTION

Natural chlorophyll- and retinal-containing protein complexes (photosynthetic reaction centers, bacteriorhodopsin) perform with high-quantum and energetic efficiencies with light-dependent transfers of charges, electrons, and protons.

Experience gained in studies of their functional, structural, and dynamic organization by kinetic optical spectroscopy, potentiometry, radio and Mossbauer spectroscopy, scanning tunneling microscopy, and small-angle scattering X-ray analysis allows one to propose membrane proteins, native or modified (including those produced by mutagenesis), to use in biotechnology and microelectronics.

The unique combination of photochromic, photoelectrogenic, and electro-optical properties per se has generated the interest of investigators from the viewpoint of fundamental science, and as a promise in the perspective of using them in new ultrafast data processing and data storage systems as well as in dynamic pattern recognition [1–3].

2. GENERAL INFORMATION ON THE STRUCTURAL–FUNCTIONAL ORGANIZATION OF BACTERIORHODOPSIN AND REACTION CENTERS

2.1 Bacteriorhodopsin

Bacteriorhodopsin (BR) is a pigment-protein complex of the purple membrane (PM) of the *Halobacterium salinarium* (formerly *halobium*) cells. It functions as a light-driven proton pump [4,5]. BR represents a single polypeptide chain of 248

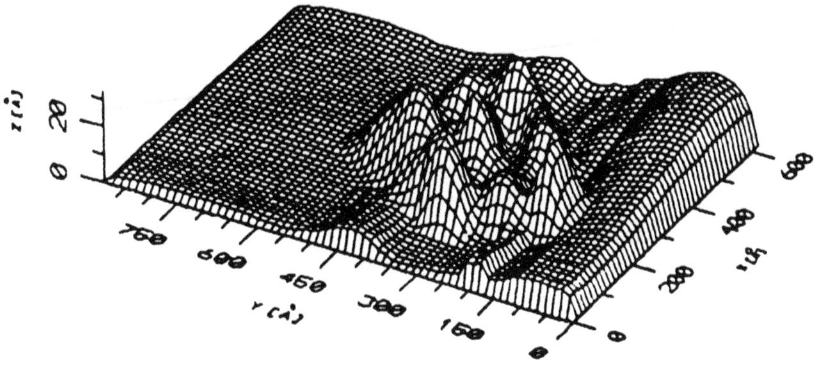

Figure 1. STM image of a PM fragment on a substrate plate. A drop (0.3 mkl of PM fragments av. o.d. of about 300 Å) produced by ultrasonic disruption of BR-containing patches was spread on a current-conductive support of pyrolytic graphite. The tungsten needle was used as a scanning tip. Measurements were taken at constant tunneling current I_t=0.3–0.7 nA. The voltage across the tip and the sample was V_t=0.3–0.4 V with the positive potential applied to the substrate. The scanning rate was no more than 200 Ås^{-1}. Temp. approx. 20 °C, relative humidity approx. 40%.

amino acid residues arranged in seven transmembrane α-helixes. Retinal is the BR chromophore. It binds to lysine-216 via protonated Schiff base linkage, forming a center responsible for its absorption at about 570 nm. The native PM (a patch) consists of a two-dimensional hexagonal lattice of trimer clusters of BR in a lipid bilayer [6]. A part of this structure is visualized in Figure 1.

In spite of the widespread view about poor electron conductivity of most biological membranes and proteins [7], we were probably the first [8] to succeed in imaging PM fragments by the method of scanning tunneling microscopy (STM). One can see a distinct regular hexagonal structure composed of individual "bumps," each obviously being a trimer of BR molecules. A mean distance between the clusters is about 100 Å; that is somewhat larger than the estimates obtained by other methods (62 Å in the native PM). It may be the result of spreading the BR molecules over the graphite support due to van der Waals forces. The pressure gradient can somewhat deform the protein and lipid membrane components. This may also be a reason for a slightly smaller height of the BR clusters (ca. 30 Å compared with ca. 45 Å in the native PM).

It is apparent that in constructing biotechnical—namely photoelectric devices using PMs—one has to take into consideration the balance of the conductive and the insulative properties of the PM films as a function of temperature, humidity, and other parameters, as well as its packing geometry on the substrate surface. Equally, this concerns RC preparations and we shall touch upon this aspect below.

As for the mechanism of protein conductivity and image generation, we suggested earlier [9]: (1) electron tunneling over large distances via the conducting channels formed by elements of the structure of the pigment-protein complex, mainly through molecular groups with a developed system of conjugated π-bonds; (2) proton-type conductivity of hydrated proteins—the charge is transferred in this case via a system of hydrogen bonds coordinated by polar hydration centers. The latter now seems to us the more probable, taking into account the new data of Guckenberger et al. [10].

After light absorption by the BR chromophoric center (maximum of the main band, ca. 570 nm) and *trans*- to *cis*-isomerization of retinal, the pigment-protein complex undergoes cyclic conversions at the expense of photon energy stored with the regeneration of the initial BR all-*trans* form. The photocycle consists of a sequence of thermal transitions via a number of intermediates, designated, as it is generally accepted by Latin letters:

$$BR_{570} \text{->} K_{610} \text{->} L_{550} \text{->} M_{412} \text{->} N_{560} \text{->} O_{640} \text{->} BR_{570}$$

The intermediates differ certainly by their isomerization state of retinal, absorption spectra, life-times in the subpico–millisecond range and the protonation state of the Schiff base.

The most significant "key" functional state is attained in the so-called M form, absorbing approximately at 412 nm (see number in subscript, the same meaning for all intermediates). The formation and the decay of this form corresponds to the deprotonation and the reprotonation of the Schiff base. In that way cyclic BR interconversion leads to vectorial intraprotein (i.e. through the PM) proton transfer [11–12].

Now, in spite of the absence of the X-ray diffraction analysis of BR crystals, the structure of the protein interior is extensively studied by various other methods. The locations of the retinal, its binding site, and amino acid residues that relate to the light-driven proton pumping are well known [6]. It is established that the major role is played by the aspartate 85, in which is involved in the proton release from BR and aspartate 96, and which is crucial for the reprotonation of the Schiff base [12]. So the proton movement seems to be a series of translocations between amino acids serving as proton donors and acceptors. Water molecules also take part in this process forming inputting and outputting proton semichannels of BR.

In the framework of this presentation it is important to emphasize the following.

1. Since the regeneration of initial BR from the photoinduced state K is temperature-dependent, one can regulate the cycle by varying this parameter. It is actually possible to stop the process at the desired spectrally and kinetically resolved stage.

2. From every BR state fixed (e.g., by temperature lowering or by some other means, see below) it is possible practically to return back to the initial BR by photoactivation of the corresponding intermediate in its specific absorption band [13].

As will be seen, further photoreversibility of selected reactions, mainly BR$\leftrightarrow$M, is the essential factor in the development of some biotechnical devices based on BR.

2.2 Reaction Centers

The primary process of photosynthesis, as it is known, is characterized by a high efficiency, both quantum (up to 100%) and energetic (of about 60%), of light energy transformation into an electrochemical potential of ion-radical products. This occurs in so-called photosynthetic reaction centers (RC)—molecular complexes of porphyrine (namely chlorin) pigments and redox cofactors of quinone nature incorporated into the structure of the specific membrane proteins. To date, considerable experience has been gained in fractionating photosynthetic membranes—in particular, from some bacterial species. The smallest particles that retain light-driven electron transfer activity can be extracted by treatment of membranes with special detergents. The rather typical bacterial RC contains three polypeptides, designated as light (L), medium (M), and heavy (H) subunits. To the L and M subunits are four noncovalently bound molecules of bacteriochlorophyll, two molecules of bacteriopheophytin, two quinone-type molecules, and one iron atom. The stoichiometry of the polypeptides, chlorin pigments, and redox cofactors, complete amino acid composition, and primary structure of the protein subunits are now well known. L, M, and H subunits consist, respectively, of 281, 307, and 260 amino acid residues; the general molecular mass of the complex is about 80 kDa. The problem of crystallization of this highly hydrophobic membrane protein was first solved by German scientists, who won a Nobel prize [14]).

For the same reasons that were mentioned in the case of BR we will present here the STM images of a single macromolecule of the most typical bacterial RC, which were also first obtained by us at Moscow University [9] (crystallographic data are also now available for this type of RC from *Rhodobacter sphaeroides* [15]).

One can see from Figure 2 that the molecular image of the RC is "pancake-shaped"; that is, it is somewhat flattened with a length of about 105 Å and height of about 15 Å. The difference of about 30% from corresponding crystallographic sizes, as in the case of BR, also may be explained by the influence of the forces of atom–atomic interactions.

Now we shall turn to the functionality of the RC. It is well established that of four bacteriochlorophyll molecules, two (special pair or dimer) act as a photochemical electron donor and that at least one bacteriopheophytin serves to mediate charge transfer between dimer and one of the quinones, the primary electron acceptor (Q_a).

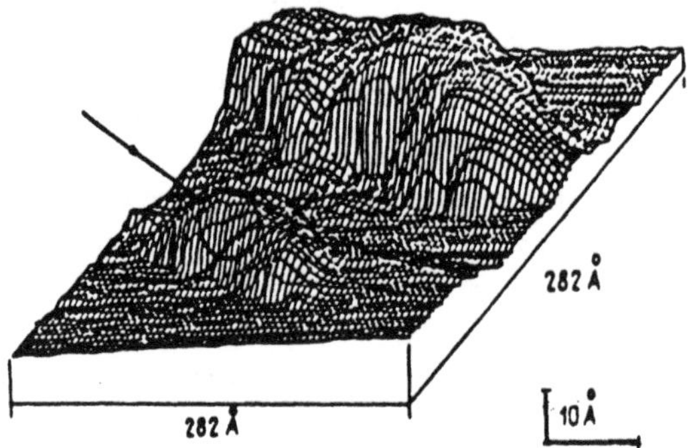

Figure 2. STM images of a separate RC macromolecule (*marked by arrow*) and clusters of these molecules. A drop of the RC preparation was placed on atomically smooth surface of highly oriented pyrolytic graphite. Stable reproducible images of RC were obtained with the tunneling voltage V_t within 0.3 to 1.0 V, the negative potential being applied to the tungsten tip. Temp. approx. 20 °C, relative humidity approx. 40%.

Upon light activation, a fast separation of the negative and positive electric charges occurs between the bacteriochlorophyll dimer and Q_a. This reaction is complete in 150 ps. Then the electron trapped on Q_a may move to the secondary quinone acceptor Q_b. This usually takes about 100 ms. Notably, the ion-radicals produced in very fast reactions are stable against recombination for about 10 ns, 80 ms, and more than 1 s at the following stages [16].

Photoelectrogenic properties of RCs may obviously be used in various biotechnical light-to-electricity converters provided the protein macromolecules can be oriented in an appropriate way.

2.3 Some Biotechnological Applications of the Photosensitive Proteins

In an approach to the problem of practical use of light energy-transducing pigment–protein complexes as nonconventional active biomaterial for molecular electronics, BR shows much promise. By the Langmuir–Blodgett–Schaefer technique or specially elaborated method of electro-sedimentation, highly ordered membrane structures containing this protein can be produced as mono- and multi-layer thin films or as polymer matrixes deposited on solid, optically transparent current-conductive technological supports. Incorporated in such systems, BR shows good adhesive properties and compatibility with electronic components. Certain other important properties of BR are:

- molecular sizes, a property valuable in term of miniaturizations of optico-electronic systems and highly dense packing of active elements;

- biosynthesis (ecologically pure and rather cheap process) according to the preset genetic program in which provisions can be made for the reproduction of identical structures;
- quick response (even in the picosecond time domain) and reversibility—a property required in phototransducers—and sufficient spectral selectivity; and
- relatively high resistance to photochemical and thermal destructive effects; in dry films the functional activity of BR (capability of performing chromatic transitions and electrogenesis persists for a few years) (see [1-3] for more extensive description of BR as an advanced material for optics and microelectronics).

In recent years a number of bioelectronic devices based mainly on BR have been described in the literature, aimed at solving various engineering problems. Here we shall focus on these studies where BR is used only as the photochromic material or the nonlinear optics media. Applications using its photoelectric properties have been discussed at length in our previous review on photobiosensors [2].

In nonlinear optical systems, BR was investigated as an environment for second-harmonic generation to be used in optical modulators and switches [17, 18].

The potentiality of BR as material for biochips and biocomputers has been discussed [19]. Today, however, prospects of its use seem remote.

Attempts have been made to record optical information photographically (irreversible) using chemically modified BR [20, 21]. Some Japanese firms, including CANON, finance studies along these lines.

Fairly high changes in the absorbance and refractance of BR matrix make it attractive as a material for holographic recording of optical information in a reflected and a transmitted light flux, a reason for activating such studies [3, 22].

Chen and Lewis [23] worked out an ultrafast high-capacity storage device on the basis of BR immobilized in a polymer matrix. The nonlinear properties of BR were used to build up a three-dimensional writing and nondestructive reading of optical information; namely, two-photon absorption of light in the main absorption band is used for information writing and the second-harmonic of laser radiation beyond this band is used for information reading. BR with retinal substituted for its analog was proposed to be used for the improvement of the characteristics of the system. The point is that the substitution of BR retinal changes the active spectral band of the BR preparation. The exchange of retinal for an azulene analog causes a shift of the absorption band toward the near-infrared region [24]. The use of 13-trifluoromethylretinal results in a pigment the absorption band of which resembles the radiation bands of the semiconductive ejection lasers [25].

Little is known to date about any biotechnical applications of the bacterial RCs in optics and microelectronics at the high levels already achieved for BR. Meanwhile it has been established due to the intensive research in various labs from the early 1980s up to now [26-31] that RCs can form stable monolayers at the air–water interface and be transferred on various supports with preservation of their specific photoelectron transferase activity in multilayer arrays. Our recent advances in this field of investigation were aimed not only at a better understanding of the molecular mechanism of light-driven charge transfer but also at the design of bioelectronic devices containing RCs as an active functional component.

3.   AUTHOR'S CONTRIBUTIONS TO PHOTOSENSITIVE DEVICES BASED ON BR AND RC

We shall now turn to some variants of biotechnical systems that have already been created by us and co-workers at Moscow University.

Figure 3 schematically represents the principle of disk carriers of optical information (digital memory) based on BR as an active material and operating in a "record-read-erase" mode in combination with a programmable laser scanner to provide random-access capability to data location.

The photochromic environment—BR in a gelatin film with good homogeneity and flatness—was prepared by a polymer matrix spin-coating technique. In the cross section at the top of Figure 3 one can see informational tracks (positions 6) that are native BR, and dividing tracks (positions 7) produced by laser burning of the pigment (see [32] for more technological details).

We used the BR↔M reversible phototransformations (transitions with good photochromic characteristics [1]). Optical information can be recorded on the memory disk by green laser light, either pulsed or continuous. A change of a color occurs at the irradiated spot due to direct BR→M reaction, and one can read it with the conventional technique of absorption spectroscopy. To erase the information, blue laser light initiating M→BR conversion is used. In this case, at room temperature a rather moderate (microsecond) rate of a single bit of information storage and processing could be achieved. Using another phototransition BR↔K (see previous corresponding section), which is possible at low (<150 K) temperatures, one can make this device (optical disk) work faster, in the picosecond time domain.

The BR environment in such a device provides a high spatial resolution (about 10^3 lines per mm); an ability to reversibly record information (high cyclicity up to 10^5 record-erase cycles); long-term functioning (no less than for traditional magnetic carriers); spectral range of recording (520-590 nm) and erasing (around 410 nm); and dynamic range of 60 dB in the mode of irreversible recording and 34 dB in the reversible mode at temperature 210–215 K. The range of the recording sensitivity of typical BR films is within tens of mJ per cm^2 according to estimations made in ref. 3.

The given characteristics could certainly be improved with the use of an advanced laser scanner and optical system; the BR disk is not likely to be a limiting element.

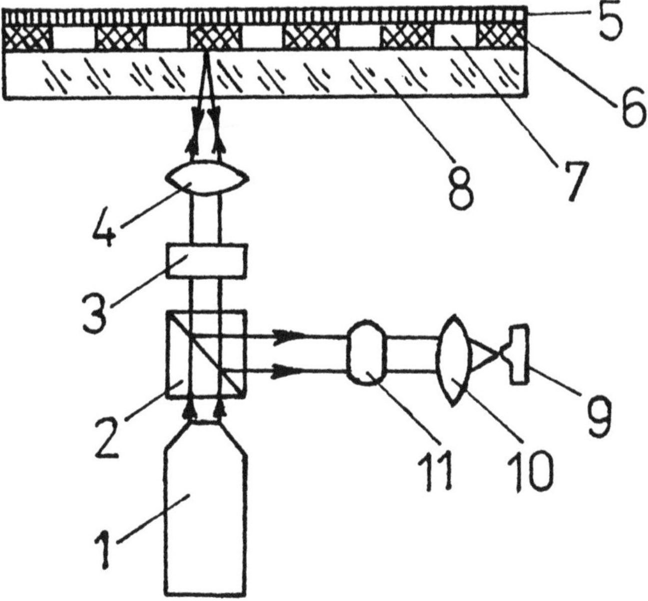

Figure 3. Disk optical information carrier based on BR with a programmable laser scanner: (**1**) Laser setup; (**2**) polarizing beam splitter; (**3**) quarter-wavelength plate; (**4,10**) objectives; (**5**) reflecting layer; (**6**) information tracks (BR film); (**7**) separating tracks; (**8**) optically transparent support; (**9**) coordinate-sensitive photodetector; (**11**) cylindrical lens.

Effective operation of optical memory disks based on BR for obvious reasons may require the M form to be stabilized. To maximize its lifetime it is known that one can use lower temperature, dehydration, chemical, or genetic modifications of BR. We have focused our attention on properties of different specific chemical modifications of BR in search for a preparation with much slower relaxation characteristics. Photochromic media with such properties may be prepared by treatment of the PM dry film with 1,2-cyclohexanedione (CHD) in the presence of borate buffer at alkaline pH. We have shown [33] that the modification by this reagent of the arginine residue Arg82 in complex with Asp85, which controls proton transfer, leads to the considerable retardation of M decay in dry films.

Figure 4 shows that after illumination of the CHD-treated PM dry film with green light (used in our optical disk for information recording) the main absorption maximum at 560 nm is bleached and a new maximum near 405 nm, due to BR→M photoconversion has appeared. It should be noted that even after 24 h in darkness more then 10% of M product still persists (curve 6). An additional blue light activation in the just photoproduced M (curve 2) absorption

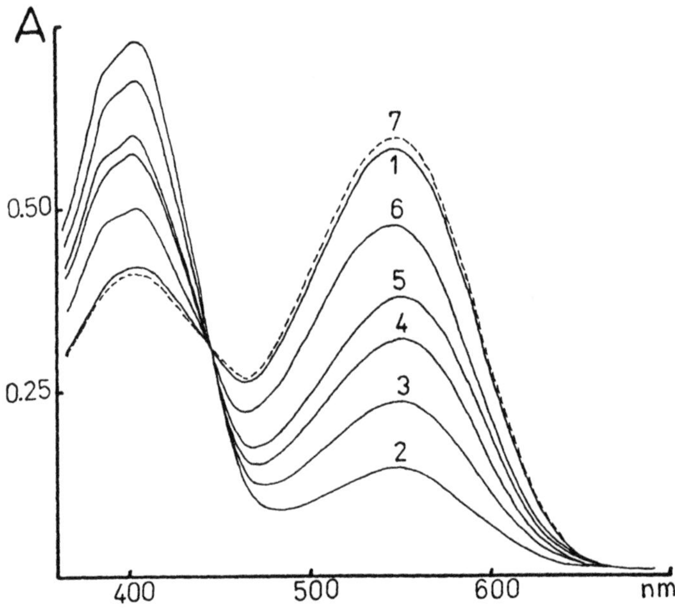

Figure 4. Absorption spectra of the CHD-treated PM film at various times of the photoinduced M intermediate conversion to the initial BR. Modified PM films were obtained by drying freshly prepared samples on a glass plate under actinic continuous illumination ($\lambda > 500$ nm, 200 W m^{-2}) after they have been soaked for 1 h in 0.5 M 1,2-cyclohexanedione in 0.1 M borate buffer, pH 8.2. (**1**) Spectrum of CHD-treated film before illumination after 48 h of dark incubation; (**2-6**) spectra immediately after green light illumination ($\lambda > 500$ nm, 200 W m^{-2}) and 10 min, 1 h, 3 h and 24 h later respectively; (**7**) spectra after green light and immediate blue light illumination ($380 < \lambda < 500$ nm, 80 W m^{-2}).

band at 380–430 nm (this light is used for recorded information erasing) causes rapid return to the initial BR state (curve 7). Thus, a new BR-based bistable photochromic material for biotechnical applications was fabricated.

In another device constructed, which utilized the excellent photochromic properties of BR, we used the physical effects first observed by us and described earlier [34-37].

In our extensive work on dry films of PMs we have found that the external field (approx. 10^7 V m^{-1}) induced bathochromic shift of the main absorption band of BR [34, 35]. In highly ordered PM films obtained by the method of electrophoretic sedimentation (see below) this effect is strongly dependent on the electric field direction [37]. The bathochromic shift under the applied electric field occurs only when field direction is from the periplasmatic to cytoplasmatic side of the membrane. We will not go into details on the origin of this bathochromic shift.

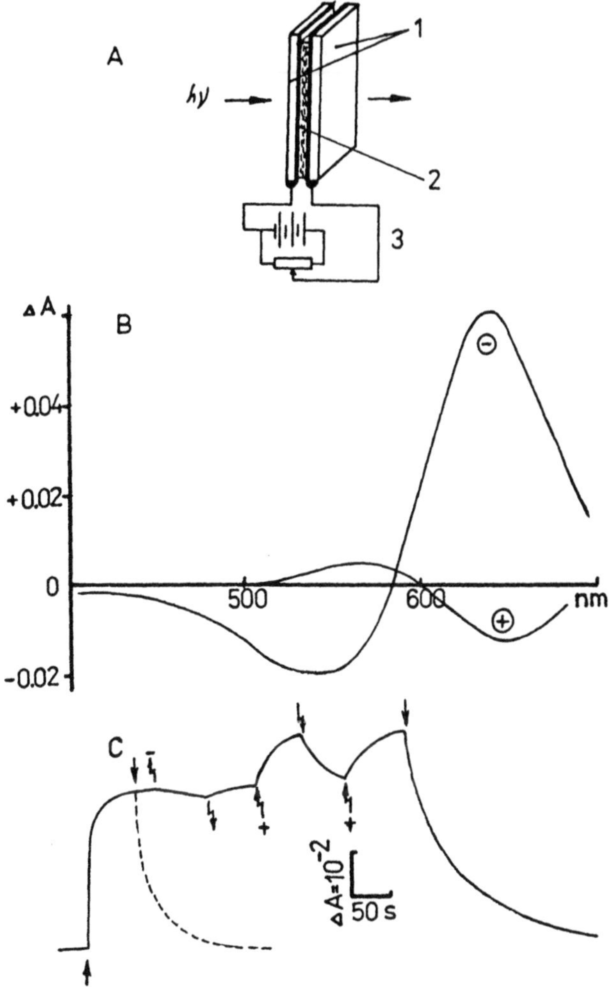

Figure 5. Electric field induced spectral-kinetic changes on highly oriented PM films. (A) Construction of the experimental cell arranged as a BR multilayer sandwiched between electrodes: (1) glass plates with semitransparent electroconductive cover (indium tin oxide); (2) oriented dry film of PMs; (3) adjustable DC source. (B) Electrically induced absorbance changes in the PM film at different direction of applied external field 10^7 V m^{-1}. (C) Kinetics of light and electrically induced absorbance changes in the PM film at 412 nm, corresponding to the M form transitions. (*Arrows up)* Light (↑) or electric field (↯) switch on; (*Arrows down*) light (↓) or electric field (↯) switch off. Sign "+"—refer to the direction of applied field from cytoplasmatic to periplasmatic side of the PM ; Sign "-" —field is applied in the opposite way.

Possible explanation of the phenomena based on the theory of electrochromism is discussed at large in our review paper [38]. Small hypsochromic shift is also observed with the field of opposite direction (Figure 5, position B). These data were confirmed later by Tsuji and Hess [39].

Here it should be stressed that BR film sandwiched between electrodes (Figure 5, position A) may be considered as an electric field-controlled photochromic element that may be of use in optical processing instruments. It is important from the technological point of view that electrically induced spectral changes of BR can be rapidly reversed by an applied opposite electric field [35]. This is especially true for the uniformly oriented PMs, which have the evident dependence on the vector of the applied field.

We are now conducting research work on the analog of BR which contains 4-keto-retinal as chromophore (maximum of the main absorbance band near 510 nm). It has been found that for this modified BR embedded in a gelatin-based polymer matrix [40] the electrically induced absorbance changes appear to be higher compared to those observed for natural BR at the same strength of the external field. Using the various analogs of BR one can obviously change the spectral range of electrically induced chromatic effects and increase its contrast.

Upon application of the external electric field in such a device (Figure 5, position A) it is also possible to regulate the kinetics of M→BR conversion in the photocycle (position C). Applying a positive potential upon the cytoplasmatic side of the PM—that is to the support plate on which the PM multilayers are formed—results in a significant prolongation of the M decay kinetics. The negative potential on this side makes this transition faster.

The effect of external electric field control of the M lifetime in films of PM was used in holographic setups [1]. The device, quite analogous to that presented in Figure 5, may serve as a voltage regulated bistable optical element, or, according to R. Birge, as a spatial light modulator with the threshold effect. He suggested that such a two-dimensional adjustable modulator could be empoyed as an optical processing environment, to modify, for example, the amplitude/intensity of spatial light distribution due to the BR↔M photochromic reactions as a function of applied electric field (voltage controls the transmittance of the photochromic film) [1].

One can utilize not only the rapid photochromism inherent to BR, but also its photoelectrogenic properties [41, 42] which relate to proton pumping capabilities. The rise and decay of the photovoltage generated by BR consists of several phases, which correlate in PM dry films to a first approximation with the transformations of the K and M intermediates. Transient photovoltage in each photocycle can certainly be used for microelectronic applications [41]. Oriented films of PM containing BR are most useful in this case [43].

Among the different methods of membrane and pigment–protein complex ordering, we shall consider only two: fabrication of films on the air-water interface by the Langmuir–Blodgett–Schaefer methods (LB), and film formation by the electro-sedimentation technique (ES).

The spreading of PM at the air–water interface yields somewhat oriented (up to 80%) monolayers (see [36] and refs. therein]. They can be then compressed and transferred on solid current-conductive optically transparent substrates. In that way, photoelectrically active multilayer assemblies can be formed. The LB technique allows superfine films to be produced containing a relatively small number of monolayers. This technique has some merits from the technological point of view. It allows the formation of hybrid superlattices using consecutive deposition of monolayers with different compositions. The electro-sedimentation technique is based on the large permanent dipole moment of PM (approx. 140 Debay per molecule and about 10^6 Debay per patch with av. o.d. 0.5 mm); the cytoplasmic side of PM charged negatively in a neutral medium. That is why an electric field of about 20 V cm^{-1} applied directly to a deionized suspension orients patches and makes them mobile. Membranes move toward the positively charged electrode (anode); the cytoplasmic side of PM orients preferentially and deposits to the anode. By means of such a technique, after electrophoretic sedimentation and subsequent drying it is possible to fabricate very highly ordered and densely packed films containing a large number of PM monolayers from a few hundred to a few thousand. Recently we have published a review paper [36] in which the major characteristics of both LB- and ES-fabricated multilayer films containing oriented PMs were compared (Table 1)

Based on ES dry films of PMs we developed a biotechnical system, that, in principle, is a spectrally selective, position-sensitive, and extremely fast reversible light–electricity transformer. Using BR as the photoactive component, it operates in the visible range, but, if realized in the same way, it may be sensitive to near-infrared with bacterial reaction centers as an active component (see below).

In the BR-based device we tried to employ such properties of this pigment–protein complex as spectral sensitivity within 200–700 nm, high quantum yield (0.64), and very quick electric photoresponse (up to several ps).

Figure 6 (position A) shows a very simplified (just to illustrate an idea of the design) photoelectric transducer model of the type mentioned above. On a transparent glass substrate plate (1) coated with a current-conductive layer of indium tin oxide (ITO) (2) ES-oriented PMs film (4) is formed (upper part in A). Such

Table 1. Comparative Properties of Langmuir–Blodgett (LB) and Electrophoretically Sedimented (ES) Films of Purple Membranes

	LB	*ES*
Number of layers	50	1000
Specific photopotential (mV per layer)	0.4	5
Specific DC resistance (Ohm cm^2 per layer)	2×10^5-10^6	5×10^7–10^{10}
Period of structure regularity (nm)*	3.6-3.8	4.64
Correlation length (nm)*	12–15	40

Notes: *As it follows from the data of low angle X-ray analysis of the preparation (see [36]   for more details).

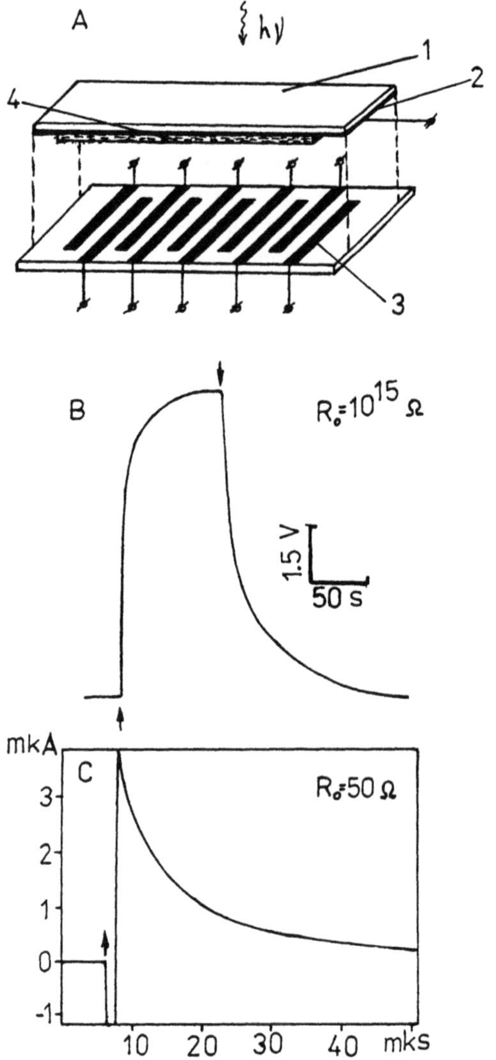

Figure 6. The spectrally selective position-sensitive photoelectric transducer based on BR and photoelectric signals in such device. Position **A:** Scheme of transducer (**1**) glass support; (**2**) indium–tin–oxid layer; (**3**) insulating plate with a metallic contact electrodes; (**4**) oriented PM film. Position **B**: Photoelectric response under continuous light illumination in a voltage mode, measured by electrometer VA-J-51 (Germany) with 10^{15} Ohm input resistance. Position **C**: Laser-induced photoelectric response measured in the current mode by op-amplifier OPA600BM (Burr-Brown Corp., Tucson, AZ, USA) with input resistance 50 Ohm.

films usually have properties (absorption spectrum, light-induced absorbance difference spectrum, decay kinetics of the M intermediate) that are very similar to those seen in typical dry PM films. The other part (lower in A) of the matrix element of the phototransducer is composed of insulating supports with contact metallic (Ni) thin strips (3); in this case an array of 10 flat elements are divided one from each other. To be functionally active, the upper and lower parts of the phototransducer (Figure 6, A) must be screwed tightly "face-to-face" to provide close contact with the oriented BR multilayer and metallic strips. Thus the oriented PM film becomes imposed between the transparent conductive substrate and array of metallic elements, each having a tap for wiring in measurements (3). Photoelectric response of this transducer with the structure, ITO/ES-BR/Ni, may be recorded in voltage or current modes depending on the input impedance of the amplifier used. The typical kinetics of photovoltage ($R_o = 10^{15}$ Ohm) and photo-current ($R_o = 50$ Ohm) rise and decay are depicted in Figure 6 at positions B and C, respectively. It is worth noting that the amplitudes of photoelectric responses from the uniformly oriented PM films, especially for ES–BR multilayers, greatly exceed (at least two orders of magnitude) that recorded for the randomly oriented PMs [44]. It is also important for microelectronic applications that, as in the case of photochromic changes of oriented PMs, the photoresponses of BR can be easily stabilized in time by temperature lowering and rapidly reversed at any temperature by additional light illumination [45].

The electrodes (low part of the position A in Figure 6) certainly may be arranged in different ways and constructed from different materials, preferentially of noble metals (Au, Pt, etc.).

Below we will present some experimental measurements of photovoltage generation due to charge displacements in activated BR molecules taken from the more advanced matrix element which consisted of two rows of 12 1 mm x 1 mm flat metallic elements. An electrometer (VA-J-51, Germany) with $R_o=10^{15}$ Ohm was used to measure the photovoltaic signal under continuous illumination ($\lambda>500$ nm, 200 W m^{-2}). The investigation of this model device yielded the following photoelectric characteristics: (1) the photovoltage light-dependence in the range of intensities down to 200 W m^{-2} represented the linear range; (2) the voltages measured across each of the individual elements were virtually the same (maximum about 0.7 V) for uniform illumination of the elements in the phototransducer; and (3) the spread of photovoltages was within 10%. This indicated the functional homogeneity of the light-sensitive surface of the sensor. With the light beam focused upon the edge of the matrix, we observed different photoresponses of the constituent element of the array. The ratio of the signals generated by the brightest and darkest area of the matrix was rather high (the exact value depended upon the variant of the construction of the photoelectric transducer). The implication is that by improving the optical excitation and measuring systems the development of a spectrally selective and position-sensitive device based on the above principle

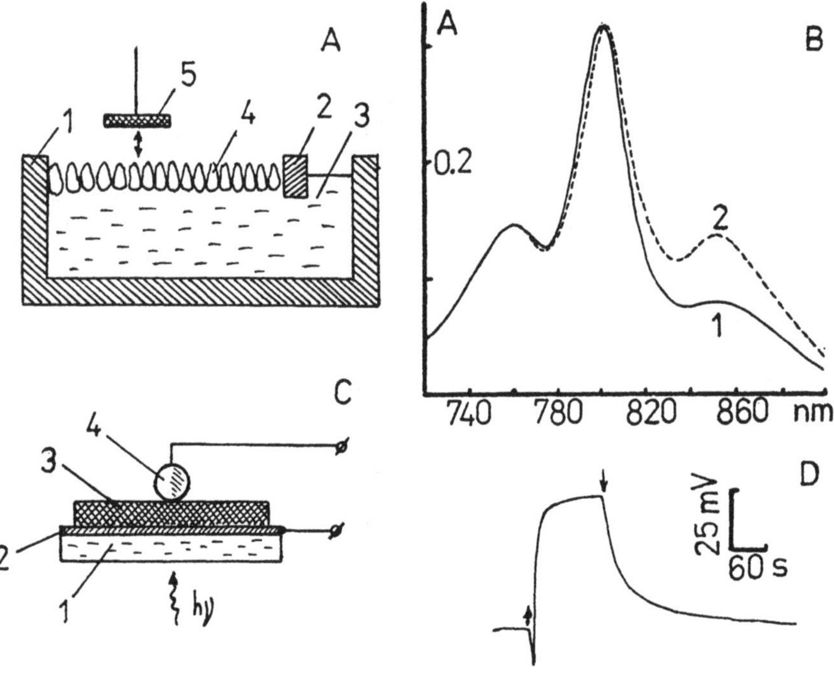

Figure 7. Fabrication of the Langmuir film of the isolated reaction centers from *Rhodobacter sphaeroides* and its photoelectric response measurements. Position **A**: (**1**) Langmuir trough; (**2**) surface barrier; (**3**) water subphase (3 mM Na-phosphate buffer, pH 6.8); (**4**) RC monolayer on the air–water interface; (**5**) electroconductive support (glass with indium–tin–oxide layer) on a dipping driver for Langmuir–Schaefer deposition. Position **B**: (**1**) Absorbance spectrum of glass plate coated with 30 RC monolayers imposed between 5 monolayers of fatty acid (myristic), the spectrum being measured immediately after deposition by Langmuir-Schaefer method at constant surface pressure 30 mN m^{-1}; (**2**) absorbance spectrum of the same sample submerged in solution of dichlorophenol-indophenol with ascorbate (10^{-4} M) for 3 min and dried in the air. Position **C**: The scheme of photopotential measurements. (**1**) glass plate; (**2**) elctroconductive layer (indium–tin–oxide); (**3**) oriented Langmuir RC film; (**4**) drop of In-Ga paste as a second electrode. Electrometer with input resistance 10^{15} Ohm. Position **D**: Photo-induced electrical response of a multilayer film composed of 30 RC monolayers between of 5 monolayers of fatty acid under continuous illumination ($\lambda > 680$ nm, 900 W m^{-2}).

is really possible. An important problem is to create a well-designed multi-channel analyzer for readout of the photoelectric responses from individual elements in series or in parallel.

Applications for position-sensitive devices may embrace many areas of science and engineering concerned with data recording, data processing, and data transmission using scanning/research means, control, measuring and other systems.

At the end of this section we will turn again to the RCs.

The absorption spectrum of the RCs Langmuir film thus obtained (Figure 7, position B, 1) coincides well with that of the RC suspension. The only discrepancy is a small shift in the position of the maximum of the photoactive bacteriochlorophyll dimer: 850 nm for the Langmuir film (the same was observed in [26–28]) and 865 nm for the RC suspension. It is obvious that the dehydration of RC's protein in the film accounts for this because moistening causes the peak to shift toward the longer wavelengths. The commonly observed ratio of 1:2:1 for the peaks at 760, 800, and 865 nm seen in a multilayer film made of a RC suspension is different for the Langmuir film. This is because the longwave absorbance of the dimer is noticeably smaller in these nultilayers. However, the extinction of the longwave maximum recovered to its normal level after the Langmuir film was kept for 3 min in a solution of dichlorophenol–indophenol (10^{-4} M) reduced by excess sodium ascorbate and dried in the air (position B, 2). The dark recovery kinetics of photoxidized dimer in this multilayer RC film following a single light flash appeared to be composed of two kinetic components with lifetimes of approximately 100 ms and a few seconds. The contributions and lifetimes of these components are nearly the same for Langmuir film and dry film of nonoriented RCs for given room humidity (35–40%). This indicates that RC protein in LB-film retains its native photochemical properties. The activity decreased by no more than 15% after the film was stored at +4 °C for 3 months. The estimated packing density yielded a value of 2.2×10^{13} RC per cm^2. It corresponds to the area occupied by one molecule 4.5×10^2 $Å^2$. This value is the same as that estimated by X-ray analysis [15].

After constructing a photovoltage measuring cell on the base of such a sample (Figure 7, position C) a typical photoelectric response as shown in Position D can be measured. The sign of photopotential (plus on the glass support) provides evidence that the H-subunit of RC protein in monolayer is oriented preferentially at the air–water interface toward the water subphase. According to our results, the ratio for oppositely oriented RC is about 60:40, which is in agreement with the data of Tiede [26, 27]. However, in new and very thorough investigations of Alegria and Dutton [28] it was found that the degree of orientation of RC protein in LB-film depends on many factors, such as type of detergent, content of subphase, concentration of lipids, and method of film deposition on the support. In the paper of Yasuda et al. [46], a very high degree of orientation of RC in Langmuir monolayer was achieved (80:20), as was determined by the ELISA (Enzyme Linked Immuno Sorbent Assay) method. Recently, Alegria and

Dutton have reported at a Gordon's Conference on Photosynthesis (1991) that they successfully applied a method of electro-sedimentation for producing highly oriented films from phospholipid vesicles in which RCs were embedded. However, more versatile studies are still needed to evaluate the possibility of using RCs as a material for bioengineering devices.

4. SOME PERSPECTIVES

Let us give two examples of possible non-traditional uses of the light-sensitive pigment-proteins.

First, it is known that in bacterial RCs, as well as in photosystem 2 RC complexes of higher plants, the light-induced electron transport is mediated by quinone cofactors (Q_b). This pathway can be blocked by some pesticides. Natural quinone appears to be substituted by the pesticide which attaches tightly to the site of Q_b binding. The blockage of electron transport may be observed by conventional spectroscopic methods by which the functional activity of the RC is monitored. It is, however, possible to increase the sensitivity of detection of toxicants. This is important in the testing of drinking water and the quality of foods. In fact, the maximum admissible content of the pesticide, atrazine, which is most widely used in Western countries, is 0.1 mkg l^{-1}. Recently German scientists (Prof. R. D. Schmid, Centre of Biotechnological Research, Braunschweig, FRG) during the International Symposium on Biosensors (Moscow, 1992) introduced an interesting idea of an assay with chemical modification of Q_b, expelled from RC interior by atrazine and its subsequent bioluminescent determination. It is supposed that in this way pesticide activity can be monitored very effectively by sensors using bacterial RC.

Second, using photoelectrogenic properties of the RCs and BR, new highly-sensitive biosensors can be developed in which the classical potentiometric sensors, modified to be light-addressable, and selective field-effect transistors can be advantageously combined. An approach to the problem rests upon the possibility of incorporating the RC and BR, alone or together, with other specific sensitive compounds into the structure of highly ordered densely packed layers. By use of the Langmuir–Blodgett–Schaefer technique it is possible to produce fairly stable mono- and multilayer films, including ensembles based on polymer matrices which can be deposited on supports made of different materials or on surfaces of microelectronic devices. Such biosensors, provided with appropriate optics and electronics, can be used to detect efficiently different ions and chemical compounds, or to identify various molecules in environmental monitoring.

REFERENCES

[1] Birge, R.R. (1990). Photophysics and molecular electronic applications of rhodopsins. *Annu. Rev. Phys. Chem.*, **41**, 683-733.

[2] Chamorovsky, S.K., Kononenko, A.A., Lukashev, E.P. (1990). Photosensitive biosensors. In Biotechnology (*Achievements of Science and Techniques*, (Egorov, A.M., Ed.), Vol. 26, pp.76-133. VINITI, Moscow. (in Russian).

[3] Oesterhelt, D., Brauchle, C., Hampp, N. (1991) Bacteriorhodopsin: a biological material for information processing. *Quart. Rev. Biophys.*, **24**, 425-478.

[4] Oesterhelt, D., Stoeckenius, W. (1971). Rhodopsin-like proteins from the purple membrane of Halobacterium halobium. *Nature New Biol.*, **233**,149-154.

[5] Birge, R.R. (1989). Nature of primary photochemical events in rhodopsin and bacteriorhodopsin. *Biochim. Biophys. Acta*, **106**, 293-327.

[6] Henderson, R., Baldwin, J.M., Ceska, T.A., Zemlin, F., Beckmann, E., Downing, K.H. (1990) Model for the structure of bacteriorhodopsin based on high-resolution electron cryo-microscopy. *J. Mol. Biol.*, **213**, 899-929.

[7] Hansma, P.K., Elings, V.B., Marti, O., Bracker, C.E. (1988). Scanning tunneling microscopy and atomic force microscopy: application to biology and technology. *Science*, **242**, 209-216.

[8] Kononenko, A.A., Lukashev, E.P., Panov, V.I., Fedorov, E.A. (1990). Scanning tunneling microscopy of functionally active bacteriorhodopsin-containing membrane fragments from Halobacteria. *Reports of USSR Acad. Sci.*, **315**, 1252-1255 (in Russian).

[9] Alekperov, S.D., Vasiljev, S.I., Kononenko, A.A., Lukashev, E.P., Panov, V.I., Semenov A.E. (1989). Scanning tunneling microscopy of photosynthetic reaction centers. Chem. Phys. *Lett.*, **164**, 151-154.

[10] Guckenberger, R., Wiegrabe, W., Hillebrand, A., Hartmann, H., Wang, Z., Baumeister, W. (1989). Scanning tunneling microscopy of a hydrated bacterial surface protein. *Ultramicroscopy*, **31**, 327-332.

[11] Lozier, R.H., Bogomolni, R.A. and Stoeckenius, W. (1975). Bacteriorhodopsin: a light-driven proton pump in *Halobacterium halobium. Biophys. J.*, **15**, 955-962.

[12] Butt, H.J., Fendler, K., Bamberg, E., Tittor, J., et al. (1989). Aspartic acids 96 and 85 play a central role in the function of bacteriorhodopsin as a proton pump. *EMBO J.*, **8**, 1657-1663.

[13] Balashov, S.P., Litvin, F.F. and Sineshchekov, V.A. (1988). Photochemical processes of light energy transformation in bacteriorhodopsin. In: *Physicochemical Biology Reviews*, (Skulachev, V.P., Ed.), Vol. 8), pp. 1-61. Harwood Academic Publishers GmbH, Chur, Switzerland..

[14] Deisenhofer, J. and Michel, H. (1989). The photosynthetic reaction centre from the purple bacterium *Rhodopseudomonas viridis. EMBO J.,* **8**, 2149-2170.

[15] Allen, J.P., Feher, G., Yeates, T.O., Komiya, H., et al. (1987). Structure of the reaction center from *Rhodobacter sphaeroides* R-26: the protein subunits. *Proc. Natl. Acad. Sci. USA*, **84**, 6162-6166.

[16] Kononenko, A.A., Rubin, A.B. (1986). Bacterial photosynthesis: protein dynamics and charges transfer. *J. All-Union D.I. Mendeleev Chem. Soc.*, **31**, 502-514. (in Russian).

[17] Aktsypetrov, O.A., Akhmediev, N.N., Vsevolodov, N.N., Esikov, D.A. et al. (1987). Photochromism in nonlinear optics: photocontrolled generation of the 2nd harmonic by bacteriorhodopsin molecules. *Reports of USSR Acad. Sci.,* **293**, 592-594 (in Russian).

[18] Huang, Y., Chen, Z. Lewis, A. (1989). Second-harmonic generation in purple membrane-poly(vinyl alcohol) films: probing the dipolar characteristics of the bacteriorhodopsin chromophore in bR_{570} and M_{412}. *J Phys. Chem.*, **93**, 3314-3320.

[19] Inoue, K. (1987). Bacteriorhodopsin optical switches. (SANYO Electric Co.), Jpn. Kokai Tokyo Koho JP 63/231424, 3 pp.

[20] Vsevolodov, N.N., Dyukova, T.V., Druzhko, A.B., Shakhbazyan, V.Yu. (1991). Optical recording material based on bacteriorhodopsin modified with hydroxylamine. In: *Optical Memory and Neural Networks*, Mikaelian, A.L., Ed.), pp.11-20. Proc. SPIE 1621.

[21] Arai, R. et al. (1986). Protein-enzyme biochemical optical recording medium and imaging processes. (CANON K.K.), Jpn. Kokai Tokyo Koho JP 63/92947, 5 pp.

[22] Mikaelian, A.L., Salakhutdinov, V.K., Vsevolodov, N.N., Dyukova, T.V. (1991). High-capacity optical spatial switch based on reversible holograms. In: *Optical Memory and Neural Networks*, Mikaelian, A.L., Ed.), pp. 148-157. Proc. SPIE 1621.

[23] Chen, Z., Lewis, A. (1992). Nondestructive reading of three-dimensional optical data storage in bacteriorhodopsin by second harmonic generation. In: *Abstr. of the 1st World Congress for Electricity and Magnetism in Biology and Medicine*. p. 57. Lake Buena Vista, Florida, June 14-19. 1992.

[24] Asato, A.E., Li, X.Y., Mead, D., Patterson, G.M.L., et al. (1990). Azulenic retinoids and the corresponding bacteriorhodopsin analogues. Unusually red-shifted pigments. *J. Am. Chem. Soc.*, **112**, 7398-7399.

[25] Gartner, W., Oesterhelt, D., Towner, P., Hopf, H., et al. (1981). 13-Trifluoromethylretinal forms an active and far red shifted chromophore in bacteriorhodopsin. *J. Am. Chem Soc.*, **103**, 7642-7643.

[26] Tiede, D.M., Mueller, P., Dutton, P.L. (1982). Spectrophotometric and voltage clamp characterization of monolayers of bacterial photosynthetic reaction centers. *Biochim. Biophys. Acta*, **681**, 191-210.

[27] Tiede, D.M. (1985). Incorporation of membrane proteins into interfacial films: model membranes for electrical and structural characterization. *Biochim. Biophys. Acta*, **811**, 357-379.

[28] Allegria, I., Dutton, P.L. (1991). I. Langmuir-Blodgett monolayer films of bacterial photosynthetic membranes and isolated reaction centers preparation, spectrophotometric and electrochemical characterization. *Biochim. Biophys. Acta*, **1057**, 239-257.

[29] Erokhin, V.V., Feigin, L.A., Kayushina, R.L., Lvov, Yu. M., et al. (1987). Langmuir films of photosynthetic reaction centres from purple bacteria. *Studia Biophys.*, **122**, 231-236.

[30] Zaitsev, S.Yu., Kalabina, N.A., Zubov, V.P., Lukashev, E.P., et al. (1992). Monolayers of photosynthetic reaction centers of green and purple bacteria. *Thin Solid Films*, **210/211**, 723-735

[31] Erokhin, V.V., Kayushina, R.L., Dembo, A.T., Sabo, Ya., et al. (1992). Structural study of the cytochrome-containing reaction centre complex of the bacteria *Chromatium minutissimum* in solution and Langmuir-Blodgett films. *Mol. Cryst. Liq. Cryst.*, **221**, 1-6.

[32] Belchinsky, G.Ya., Bogatov, P.N., Gresko, A.P., Kononenko, A.A., et al. (1991). Application of pigment-protein complexes as carriers of optic information. *Biofizika*, **36**, 248-251 (in Russian).

[33] Lukashev, E.P., Kononenko, A.A., Abdulaev, N.G., Ebrey, T.G. (1993). Retardation of photocycle in the bacteriorhodopsin by cyclohexanedione. *Bioorgan. Chem.*, **19**, 395-405 (in Russian).

[34] Borisevitch, G.P., Lukashev, E.P., Kononenko, A.A., Rubin, A.B. (1979). Bacteriorhodopsin (bR570) batochromic band shift in the external electric field. *Biochim. Biophys. Acta*, **546**, 171-174.

[35] Lukashev, E.P., Vozary, E., Kononenko, A.A., Rubin, A.B. (1980). Electric field promotion of the bacteriorhodopsin bR570 to bR412 photoconversion in films of *Halobacterium halobium* purple membranes. *Biochim. Biophys. Acta*, **592**, 258-266.

[36] Maximychev, A.V., Kholmansky, A.S., Levin,, E.V., Rambidi, N.G., et al. (1992). Oriented purple membrane multilayers of *Halobacteria* fabricated by Langmuir-Blodgett and electrophoretic sedimentation techniques. *Edv. Mater. for Opt. and Electron.*, **1**, 105-115.

[37] Kononenko, A.A., Lukashev, E.P., Maximychev, A.V., Chamorovsky, S.K., et al. (1986). Oriented purple-membrane films as a probe for studies of the mechanism of bacteriorhodopsin functioning. I. The vectorial character of the external electric field effect on the dark state and the photocycle of bacteriorhodopsin. *Biochim. Biophys. Acta.* **850**, 162-169.

[38] Maximychev, A.V., Kononenko, A.A., Lukashev, E.P., Timashev, S.F., et al. (1988). Photochromic, electrochromic and photoelectric properties of the bacteriorhodopsin oriented films. Possible applications. *J. Phys. Chem.*, **62**, 2753-2768 (in Russian).

[39] Tsuji, K., Hess, B (1986) Electric-field induced conformational changes of bacteriorhodopsin in purple membrane films. 1. DC field effects. *Eur. Biophys. J.*, **13**, 273-280.

[40] Lukashev, E.P., Druzhko, A.B., Kononenko, A.A. (1992). Electrically induced bathochromic shift of the absorption band of 4-keto-bacteriorhodopsin in gelatine-based films. *Biofizika,* **37**, 86-90 (in Russian).

[41] Trissl, H.W. (1990). Photoelectric measurements of purple membranes. *Photochem. Photobiol.,* **51**, 793-818.

[42] Keszthelyi, L., Ormos, P. (1989). Protein electric response signals from dielectrically polarized systems. J. *Membrane Biol.,* **109**, 193-200.

[43] Varo, G., Keszthelyi, L. (1983). Photoelectric signals from dried oriented purple membranes of *Halobacterium halobium. Biophys. J.,* **43**, 47-51.

[44] Kononenko, A.A., Lukashev, E.P., Chamorovsky, S.K., Maximychev, A.V., et al. (1987). Oriented purple membrane films as a probe for studies of the mechanism of bacteriorhodopsin functioning. II. Photoelectric processes. *Biochim. Biophys. Acta,* **892**, 56-67.

[45] Lukashev, E.P., Kononenko, A.A., Rubin, A.B. (1984). Temperature dependence of the photopotential in the films of purple membranes from *Halobacteria. Biol. Membranes,* **1**, 91-98.

[46] Yasuda, Y., Hirata, Y., Sugino, H., Kumei, M., et al. (1992). Preparation of Langmuir-Blodgett films of the photosynthetic proteins. *Thin Solid Films,* **210/211**, 733-735.

INDEX